THE SOCIAL BASIS OF THE MICROELECTRONICS REVOLUTION

ALFONSO HERNÁN MOLINA

THE SOCIAL BASIS OF THE MICROELECTRONICS REVOLUTION

EDINBURGH UNIVERSITY PRESS

ACKNOWLEDGEMENTS

For this book I am indebted to many people whose advice and friendly support were important in bringing the work to fruition. In particular, I want to mention the name of Donald MacKenzie whose advice at different stages of the work was invaluable. Another person who must be named is Rosemary Goring for her conscientious editing work leading to a more readable book. Finally, I am indebted to my wife Juana, and daughters Ximena and Claudia, for all the inevitable sacrifices associated with a work of this kind, and even more so for constantly reminding me that life has a most beautiful side in the enjoyment of family life.

Edinburgh University Press
22 George Square, Edinburgh

Set in Linotron Times Roman
by Polyprint, Edinburgh, and
printed in Great Britain by
Redwood Burn Limited,
Trowbridge, Wilts

British Library Cataloguing
in Publication Data
Molina, Alfonso Hernán
The social basis of the
microelectronics revolution.
1. Microelectronics industries.
Technological innovation, 1945-1987
Social aspects
I. Title
303.4'83

ISBN 0 85224 594 7
0 85224 605 6 Pbk

CONTENTS

LIST OF FIGURES

LIST OF TABLES

ONE

INTRODUCTION

Since the world woke up to the fact that a major technological revolution, the microelectronics revolution, was on the march, a great deal has been written by scholars and policy-makers attempting to grapple with the nature, threats and promises of microelectronics technology. Most of the analyses, however, have focused on the characteristics of the technology involved and on its likely social impact. In this respect, the technology itself and its processes of development have been accepted as largely inevitable. Only its impact remains as a relevant social issue for study. Authors have examined the impact of microelectronics in areas as diverse as employment, economic performance and competitiveness, education, military developments, the Third World, and the prospects for privacy and political freedoms. A large literature of this kind now exists.

On the other hand, little attention has been given to understanding the social forces shaping the microelectronics revolution, notwithstanding the obvious importance of such knowledge for any effort aimed at controlling and directing its historical and continuing development. Not that this is surprising. Critical social analyses have always been in the minority, and all the more so with the technologies of the microrevolution, which have clearly rekindled hopes and visions of technological utopias. In addition, the very complexity of the issues involved in the historical process of social shaping of the microelectronics revolution is something that does not lend itself to easy treatment. Take as an indication the following picture suggested by two commentators, in relation to the development of the microchip alone – forget about computers or telecommunications for the moment

> Reconstructing the development of the microchip is a step on the road towards demystifying its presentation because in its past one starts to see the contours of the relations that produce it. These are revealed as neither some simple-minded social demand nor as a technological eureka, but as complex interaction of factors such as the arms' race, the battle for space, US government expenditure patterns, the budget commitments of giant corporations like IBM and AT&T alongside the fierce entrepreneurial drive of Fairchild, National Semiconductor, Intel and the like. . . This process of historical reconstruction is a difficult task, but it is essential that we make comprehensible the rapid rise of the micro by illuminating the special circumstances of its growth if there is to be any prospect of the public participating in its current and future applications. (Webster and Robin, 1981, pp320-1.)

Only a handful of analyses have come closer to explaining in these terms the social nature and dynamics of the technological changes currently taking place. But even here the approaches have been partial, dealing, for instance, with semiconductors, or computers, or electronic control, but not with the range of technologies whose electronics-based process of convergence is at the technical heart of the microelectronics revolution. Also, most approaches have tended to focus primarily on economic processes and on conflicts between workers and employers, especially surrounding automation. But, important though these issues are, a wider perspective is needed.

The general aim of this book is to shed light on the social nature of the microelectronics revolution. In particular, it seeks to put into its socio-historical context the development of the broad range of converging electronics technologies (i.e. semiconductors, computers, automatic control and telecommunications) which henceforth I shall refer to as *microtechnology*. A major concern will be to map out the major characteristics and trends dominating the current phase of development of microtechnology. But I will also explore future paths of development. In particular, I shall ask whether alternative forms of social and technological change informed by different, more humane concerns are likely to develop.

In facing these issues, the approach of this book will be to look for the explanations of technological change, primarily in the nature of, and interrelations between, those social forces whose interests have played a dominant role in shaping the development of microtechnology within the advanced societies of the capitalist world. These are the social forces which have in practice exercised the dominant control of the basic human, financial, material, time and space resources necessary for all technological processes, and which, for the same reason, have most directly influenced and shaped the development of microtechnology. In this approach, the influence of non-dominant social forces is taken as subsumed in the practical decisions and actions of the dominant social interests.

An appropriate name for this interplay of social interests shaping microtechnology is that of 'dominant social constituency' of microtechnology. This name conveys the idea that the social-shaping role of these forces is as much an inseparable component of technogical processes as is the role of machinery, knowledge, and so forth, i.e. the technical constituents of technological processes. This is relevant because in my view both social and technical factors participate equally in giving technological processes their character; they interpenetrate, and the absence of either of them would make it impossible to explain the shape of these processes as they are in practice.

The Dominant Social Constituency of a Technological Process

The dominant social forces shaping microtechnology in the post 1945 US – my focus in this book – have been 'capital', 'government', the 'military' and 'science'. As the book develops, I shall show their dominant role. At this point, however, a few preliminary remarks on the form of analysis I will use are necessary. I take as my starting point the idea that in a given society, all social forces have a social role which constitutes their *raison d'être*. The need to fulfil this *raison d'être* creates overriding interests which influence the actions of these forces. The common expression of these overriding interests is the need to ensure their access to, or control or share of, those resources which constitute the lifeblood of the specific activities which define their *raison d'être*. In practice, there is no single way by which this need is pursued or satisfied. Ways and means vary not only for different social forces and historical periods, but also for different institutions within the same social force (e.g. different companies within the social force of capital). This flexibility implies the existence of particular interests which constitute the practical expression of social forces' overriding interests. Particular interests thus relate to the more ordinary, daily activities of institutions and people. They are informed by the overriding interests of social forces, but also by other factors such as economic and political pressures, organisational traditions of institutions, legislation, quality and quantity of technical resources, and also the personalities, ambitions and visions of individuals working and making the day-to-day decisions in particular institutions.

The importance of overriding interests, however, is that they cut across this practical diversity, setting general goals and ultimate limits for the expression of particular interests. For instance, capitalist enterprises may differ in the way they pursue profit-making, but this is still their overriding interest, since profits are central to their *raison d'être*, capital accumulation. A similar situation applies to all other social forces insofar as they seek the control of, share of or access to those resources which enable their specific activities and with them the reproduction and further development of the social forces themselves. Such command of, or access to, resources may be simply reduced to the possession of effective social power, either directly by controlling economic and/or political power, or indirectly by having influence upon those in direct command of power and resources. This means that, on the basis of overriding interests, social forces may converge.

In particular, power alliances between different social forces (e.g. science and the military) may develop when the product of the activities of one force is perceived as important to the furthering of another force, and when this second force commands resources which are essential to reproduce and further the activities of the first force. In this case, the interests of the second force, say military power in the case of the military, will tend to shape the product of the activities of the first force, say the scientific and technical knowledge produced by science. In turn, by making products available to the military, science is also helping to shape the practical expression of military power. Altogether, what would have occurred is a social convergence between the forces which would advance the interests of each.

Admittedly, such a process is not straightforward. Several factors intervene, with particular interests playing an important role under given historical circumstances. Such factors include the ethics of using science to develop weapons; or external pressures such as opposition from social forces opposing a military-science convergence. Also, the existence of alternative alliances may intervene between the interests of distinct social forces such as science and the military. In looking at the process of convergence, however, my primary concern is not with the particular interests or ethical principles characterising each of these social forces nor with the internal problems and contradictions facing each of them either, as a result of their convergence or, simply, as a result of internal differences regarding issues of technological development. My primary concern is with the factors that make such convergence possible, and which, ultimately, imply that none of the opposing factors should be strong enough to prevent it. In these conditions, under favourable sociohistorical pressures or galvanising forces such as war and competition, it is possible to suggest that those social forces whose overriding interests are complementary will tend to converge into systemic interactions reproducing and advancing the interests of each.

The Second World War exemplifies the formation of a complex of social power crystallising the converging interests of government, the military, science and capital under the impetus of war. Such convergence has developed out of these forces' complementary *raison d'être* and overriding interests, which may be summed up as the pursuit of quantitative and/or qualitative accumulation of specific forms of power.[1] Note that this pursuit is not necessarily in an absolute sense. It may be in a relative sense, that is, in relation to the power of other antagonistic or competing forces. At any rate, in the case of capital this pursuit is seen in the accumulation of capital through profit-making activity, for the military in the accumulation of destructive power through improved weaponry, for science in the accumulation of scientific and technical knowledge through the advancement of the frontier of this knowledge, and for government in the accumulation of political power both nationally and internationally, through the economic, military and scientific-technical power deriving from the other forces.

True, government, the military and capital have long interacted with each other in the development of the capitalist state and, in this sense, they have never been completely separate forces. The military, for instance, is largely financed by government and, since military power is regarded as a necessary component of political power, there is a high coincidence of interests between the two. In the US, this fact is reflected in government largesse towards the military and the considerable influence of the military within government. Government interests, however, are much broader than the military's alone and they thus remain separate forces. Tension exists between these forces, with government interests sometimes acting to restrain military demands, as in the aftermath of the Vietnam War in the US. Similar conflict occurs over the interpenetration of government and corporate capital interests, where the latter's interests are constrained by government through such means as the US's antitrust laws. Yet convergence and interaction of interests remain significant.[2]

Since the Second World War, the focus for the convergence of these social forces has been microtechnology and, more generally, science-based technology which is seen as an important contributor in the accumulation of economic, sociopolitical and military power. As a result, all these forces have converged on microtechnology, creating a constituency of social interests that has shaped its development.[3]

Such a social constituency has made the necessary resources available for the advancement of microtechnology, while shaping its development according to the interests of the individual constituents. However, the relative influence of each social constituent has been neither equal nor static. Instead, it has evolved with the technological process, varying with time and circumstance. In this respect the historical pressures welding the constituents have been paramount. This theme of changing relative influence within the dominant social constituency is crucial for, as will be shown, the relative weight of each social constituent has been reflected in the development of microtechnology.

Methodological Aspects

It is important to make clear some points of methodology used in this work, particularly those relating to historical and statistical data used in reconstructing both the nature and workings of the dominant social constituency and the shaping of microtechnology by its changing interplay of interests.

First, this book refers primarily to the USA, since dominant social constituencies are mainly national phenomena. Nevertheless, I have also introduced a global and comparative perspective, touching mainly on Japan and Europe. The reason for choosing the United States is not arbitrary. It reflects, rather, the fact that the US is at the forefront of current microtechnology development and it is also the country where the most important historical developments in microtechnology have originated.

My main source of data has been the large literature — books, essays, journal and newspaper articles — surrounding the microelectronics revolution and its social constituency. I have of course had to be discriminating in my use of this, and have tried to draw only on authoritative sources. I have also sought to piece together from the disparate clues provided by this literature a coherent picture. Official statistical sources from the US and some international organisations have also provided abundant information. As regards statistical data, it is important to underline that their use is more to indicate the direction of specific trends than to provide an exact quantification of these trends. Figures can differ significantly between different statistical sources depending, for example, on definitions, methods of measurement used, etc. Such a case is the estimates of the military share of the US's research and development (R&D) expenditure which some sources put at nearly three times the amount suggested by the official figures given by the National Science Board or the Department of Commerce. When faced with such differences, as a general rule, I have made use of those data which give the less striking picture.

One limitation of basing the analysis on available data is that data required by the argument is not always readily available. Sometimes this means detours in the line of the argument so that important points can be supported with clear and systematic evidence. This book is no exception to this problem. In reconstructing the post-Second World War development of the US's social constituency of microtechnology, I found that available data, particularly statistical data depicting important historical trends, were not as rich and systematic as I would have wanted them to be. Fortunately, the same analysis could be made by looking at the research and development system in the US which has a well-developed base of statistical information. Since microtechnology constitutes only a particular facet of a much more general process involving the entire sphere of science-related technology,[4] (i.e. technology where scientific knowledge and, more generally, R&D, are basic resources inseparably related to its development), this means that the dominant social constituency of the US's R&D system is essentially the same as that of microtechnology. A clear insight into the historical unfolding of the microtechnology system, therefore, can be gained by focusing first on the history of the social constituency of the US's R&D system, the role played by its constituents, their varying interrelations, and the main trends characterising their development.[5] This is exactly what I do in this book, thus providing a solid groundwork to deal with the individual case of microtechnology, and confirming that, in essence, the same social constituency has dominated the development of both the R&D system and microtechnology in the US.

At this point, an explanation is in order regarding the content of the R&D system, especially as the science 'social constituent' refers not only to the institutions and interests of those involved in the pursuit of systematic knowledge of natural phenomena (see note), but, more broadly, to the institutions and interests of those involved in the activity of research and development. This would include science in its more common meaning as basic and applied research,[6] also all the aspects of the process of generating new technical development. In this sense, I do not depart from the content of some of the most well-known definitions of R&D, such as those of the US's National Science Foundation and the OECD Frascati Manual. According to these sources, R&D consists of basic and applied research in the sciences and in engineering and activities in development. Research, which encompasses both basic and applied, is systematic, intensive study directed toward fuller scientific knowledge of the subject studied whereas development is the systematic use of scientific knowledge directed toward the production of useful materials, devices, systems or methods, including design and development of prototypes and processes [Batelle-Columbus (1984), NSF (1984), US's Department of Commerce (1975).] In addition, basic research is experimental or theoretical work undertaken primarily to acquire new knowledge of the underlying foundations of phenomena and observable facts, without any particular application or use in view, whereas applied research is also original investigation undertaken in order to acquire new knowledge, but directed primarily toward a specific practical aim or objective (OECD, 1976).[7]

Structure of the Book

After this introduction the book is divided into six chapters, plus an appendix. Chapter 2 focuses on the US's R&D system and, as indicated above, attempts to systematise in a historical perspective the nature and role of the dominant social constituency of power which has shaped its development in the post-Second World War era. The appendix relates to this chapter and provides an analysis of the origins and development of the US's R&D system prior to the outbreak of the Second World War. It shows that a social constituency of power involving capital, government, the military and science did not arise for the first time in the US during the Second World War. In chapters 3, 4 and 5, the analysis turns to the case of microtechnology proper, showing how the above social constituency of power has effectively shaped its development in a way which has mirrored the interests, tensions and changing relative weights of the constituents under given historical circumstances. Chapter 3 specifically deals with the social shaping of microtechnology from its early development in the Second World War to about the mid 1970s. Chapters 4 and

5 identify the major characteristics, trends and issues dominating the present development of US's microtechnology.

The global nature of the industries and markets involved in this development is accounted for through a comparative perspective which compares the case of the US with that of Japan and Europe. The final pair of chapters, 6 and 7, looks at future development for society as it may arise from present trends. Here, the analysis is more theoretical and seeks to probe into the nature of the present social constituency of power, while also examining the issues involved in the emergence of an alternative social constituency whose interests put humanity at the centre of societal and technological progress.

TWO

THE SOCIAL BASIS OF THE UNITED STATES' RESEARCH AND DEVELOPMENT SYSTEM FROM THE SECOND WORLD WAR TO THE PRESENT

The United States' Research and Development (R & D) System[1] has become the focal point for the convergence of dominant social forces seeking to further themselves through its development, and who, obviously, seek to shape such a development in their own best interests. In the post-Second World War era in the United States, this convergence has involved the interests of capital, government, military and science. It is these forces which have commanded the process of social shaping of the US's R&D system; in other words, they have been its dominant social constituency.

At all times, historical pressures have deeply influenced the dominant social constituency by creating the conditions for the convergence of social forces, stimulating this convergence, legitimising it but, also, introducing conflicting pressures leading to tension among some of the constituents. In the postwar era, the most influential historical pressures have been international politico-military pressures, and national and international economic competitive pressures. Either or both have dominated the development of the US's R&D system, but with the relatively recent convergence of these presssures, strong tension has been created. This conflict was unknown when military pressures were the main underlying force in the first decades of the postwar era.

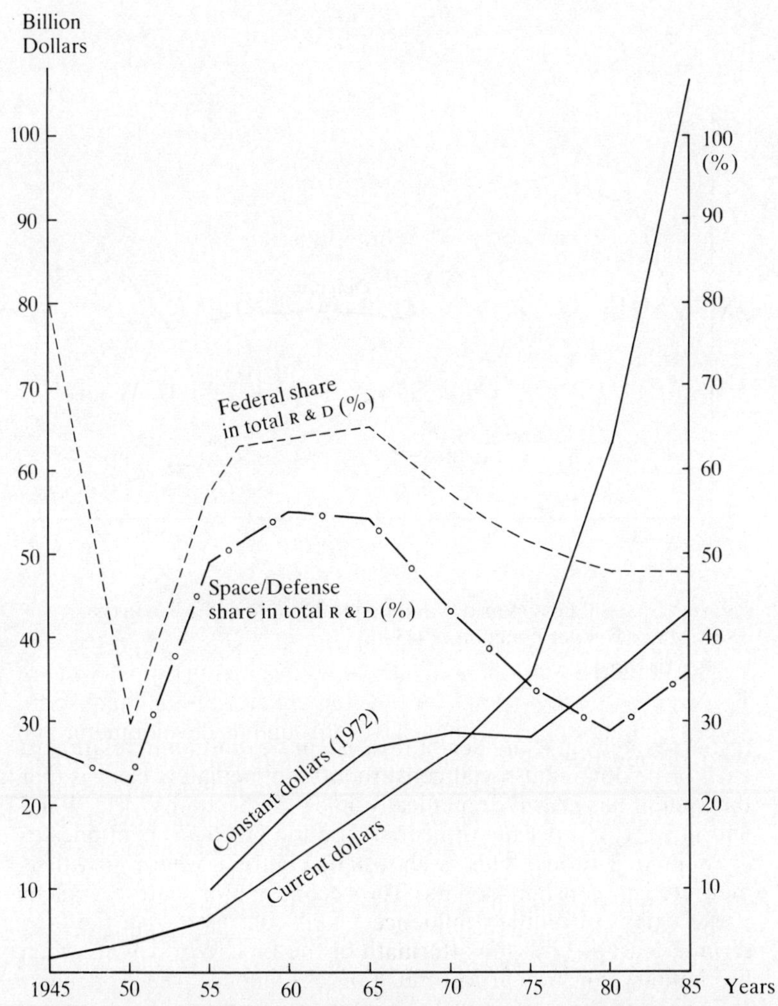

Figure 1. Post-World War II R & D Outlays in the US (1945-85).
Source. Based on figures given in Molina (1987), p.155; National Science Board (1985), *Science Indicators: The 1985 Report,* US Govt. Printing Office, Wash., DC; Department of Commerce Bureau of Census (1986). *Statistical Abstract of the United States, 1986.* Govt. Printing Office, Wash., DC

The Role of the Military

Figures 1 to 5 illustrate the variations in the R & D system since the Second World War. Focusing on the category of financial

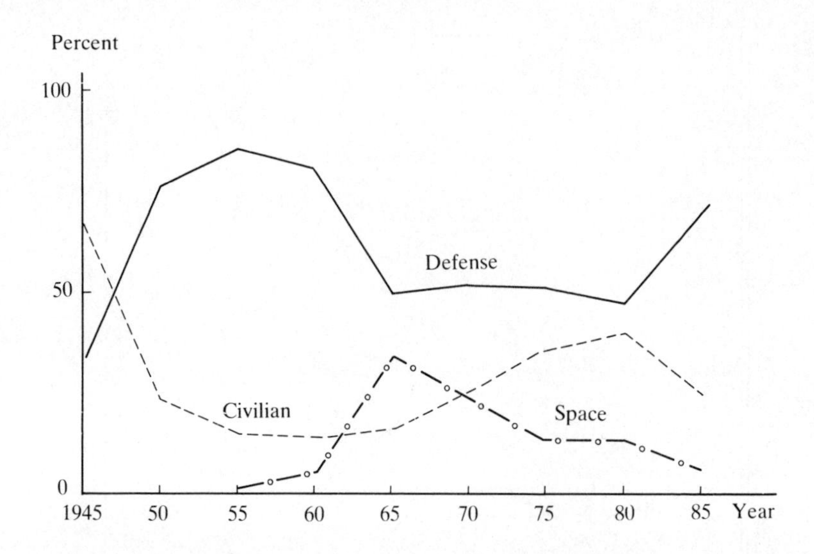

Figure 2. Share of Post-World War II Federal Funding for R & D in the US by Major Budget Function (1945-86)
Source. As in Figure 1

resources, they give an idea of the system's evolution in relation to each of its dominant social constituents. Immediately clear is that the system has grown dramatically since the Second World War,[2] and in spite of a decade of near stagnation (1965-75) continues to grow in real terms. This is shown in Figure 1 where growth is measured in constant dollars. The second major feature evident is the extent of military influence.[3] This contrasts — in the long term at least — with the aftermath of the First World War, when the military constituent lost much of its influence, see Appendix. Indeed, with the swift revival of military pressures and their endless reproduction through the Cold War,[4] the military has become a permanent dominant constituent, heavily influencing the postwar development of the US's R&D system: so much so that, for the first time on any important scale, R&D work for military purposes has come to be seen as a major source of new technologies for civilian applications: the so-called technological spin-off or overspill.[5] Thus, according to Salomon, where before military technology had relied on civilian technology, now an important reversal had taken place.

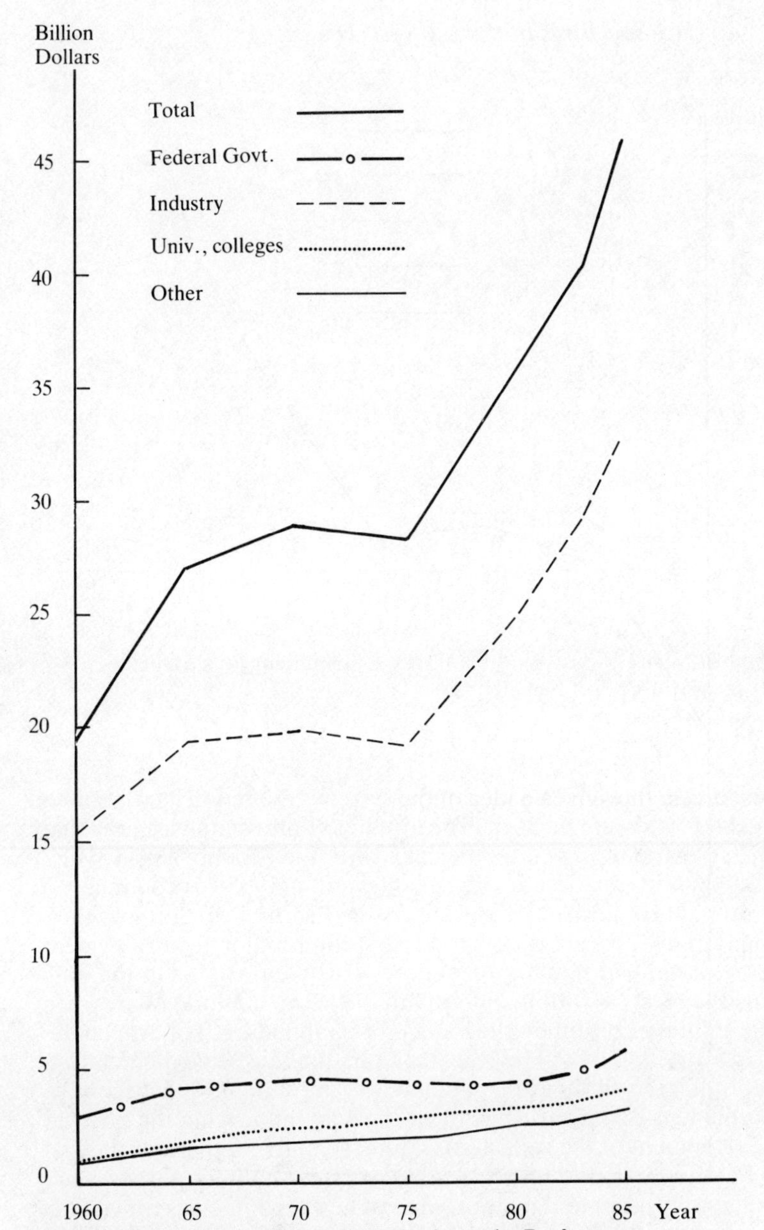

Figure 3. National R & D Expenditure in the US by Performers: 1960-85 (constant 1972 dollars)

Source. Based on Figures given in National Science Board (1985). *Science Indicators. The 1985 Report,* US Govt. Printing Office, Wash., DC

Figure 4. National R & D Expenditure in the US by Source of Funds: 1960-85 (constant 1972 dollars)
Source. As in Figure 3

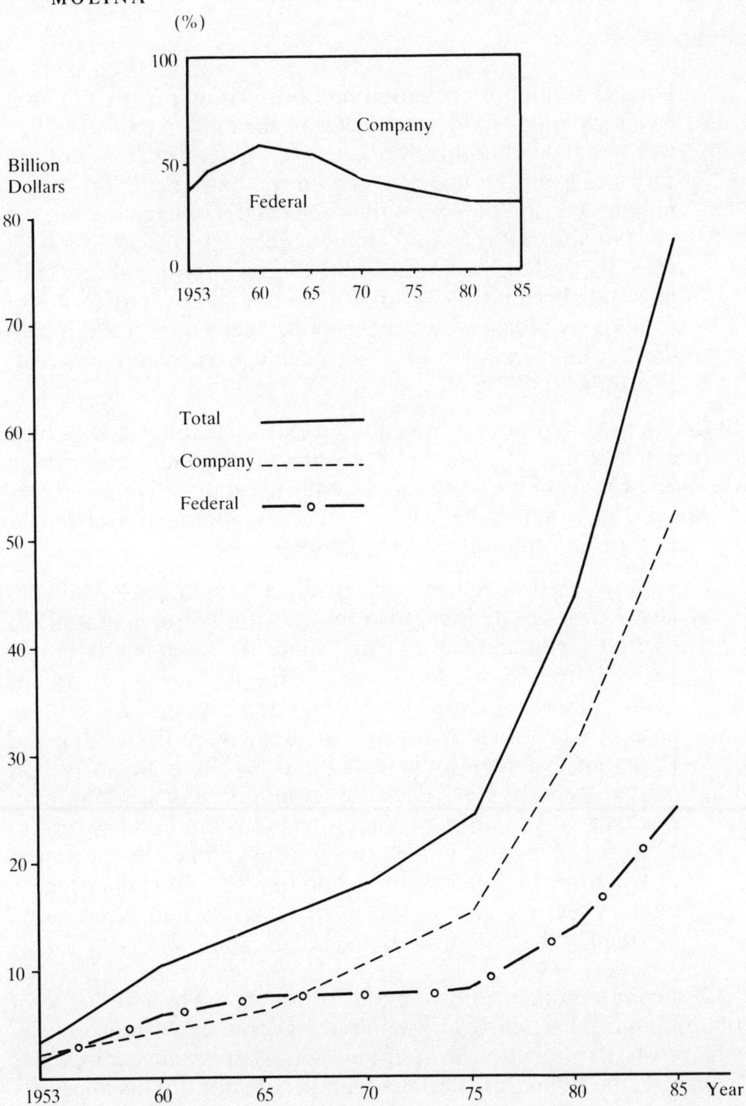

Figure 5. Post-World War II Industrial R & D Expenditure by Source of Funds (1953-85)

Source. Based on figures given in Molina (1987), p. 158; National Science Board (1985), *Science Indicators. The 1985 Report,* US Govt. Printing Office, Wash., DC

In his words,

> Until then military research had been content with adapting civil technology to the needs of war: the tank of the 1914-1918 war was indeed nothing but a cannon placed in an armoured car, and even the use of gases on the battlefield was only a military version of the progress achieved by chemical research in civil life. During and immediately after World War II scientific and technical research, conceived with military ends in mind, became the source of newly discovered forms of technology which were to be applied on a vast scale in civil life : atomic energy, radar, jet planes, DDT, computers, etc. (Salomon, 1977, p.48.)

Since then — and not surprisingly given the ongoing role played by the military — civilian spin-offs have continued, confirming the notion that military pressure is an important driving force in advancing the frontiers of civilian technology. Tirman has described the basics of the 'spin-offs' argument as follows:

> military R&D develops new 'products' for defense, but the technology created can also be transformed into something useful for commercial industry. Since the Pentagon is willing and able to underwrite high-risk 'blue-sky' research and has ample funds to give to promising ideas, it is on the cutting edge of technological advance, actually more risk-taking and adventurous than most private industry. No single institution can so steadily push back the frontiers of knowledge and discovery. And this advance inevitably finds its way into civilian commercial applications. Thus, DoD's [Department of Defense's] emphasis on technology not only enhances US military preparedness it directly benefits the civilian economy. (Tirman, 1984, pp.16-17.)

A second argument has pointed to the crucial 'market' role of the military, particularly at the initial stages of the product cycle, when costs of production are high and the risks may deter companies from investing. Thus, in relation to semiconductors for instance, 'By providing an initial market and premium prices for major advances, defense purchasers speeded their introduction into use' (Utterback and Murray, 1977, p.3). Opinions pointing to the military as a driving force in the development of postwar civilian technology have not gone unchallenged, however. In particular, the military's pervasive control over a large part of the US's R&D system, and its subsequent

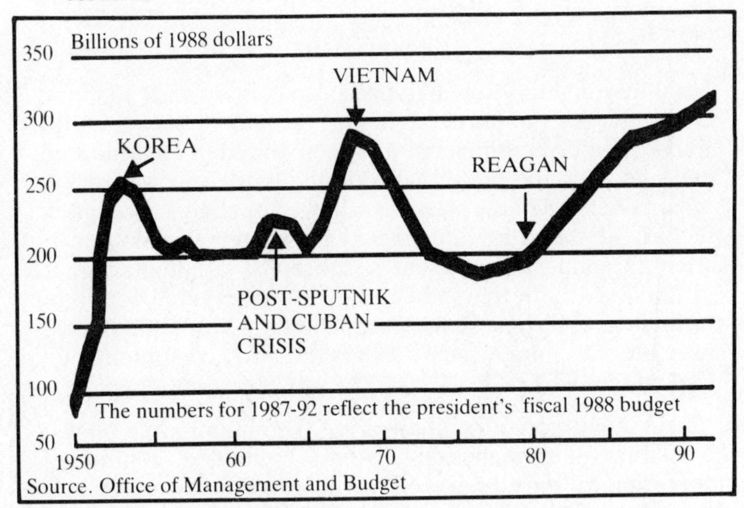

Figure 6. Total Volume of Defense Outlays in the US (1950-early 1990s). Constant 1988 dollars

Source. Financial Times 13 July 1987, p.22

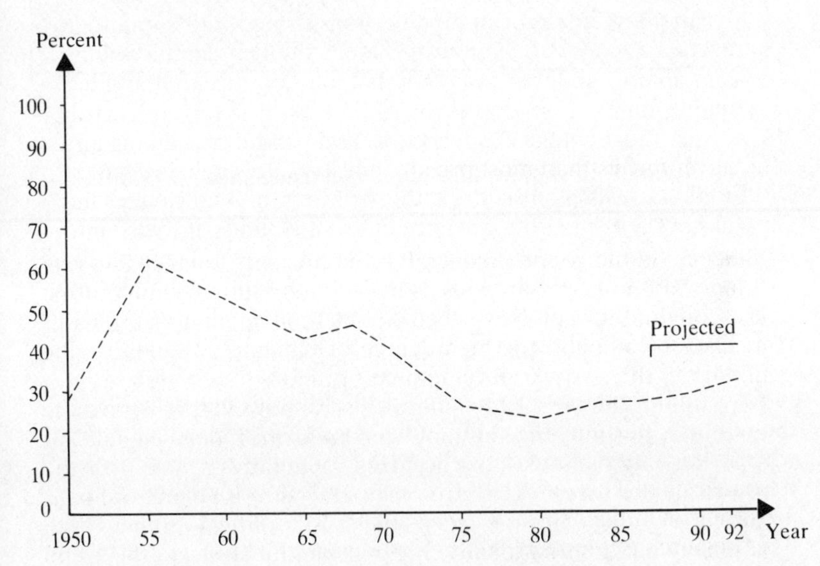

Figure 6a. Share of US Defense Outlays in Total Federal Outlays (1950-92)

Source. Based on figures given in US Department of Commerce. Bureau of Census. *Statistical Abstract of the United States 1984* and *Statistical Abstract of the United States 1986*, US Govt. Printing Office, Wash., DC, and S.A. Cain and G. Adams (1987), p.51

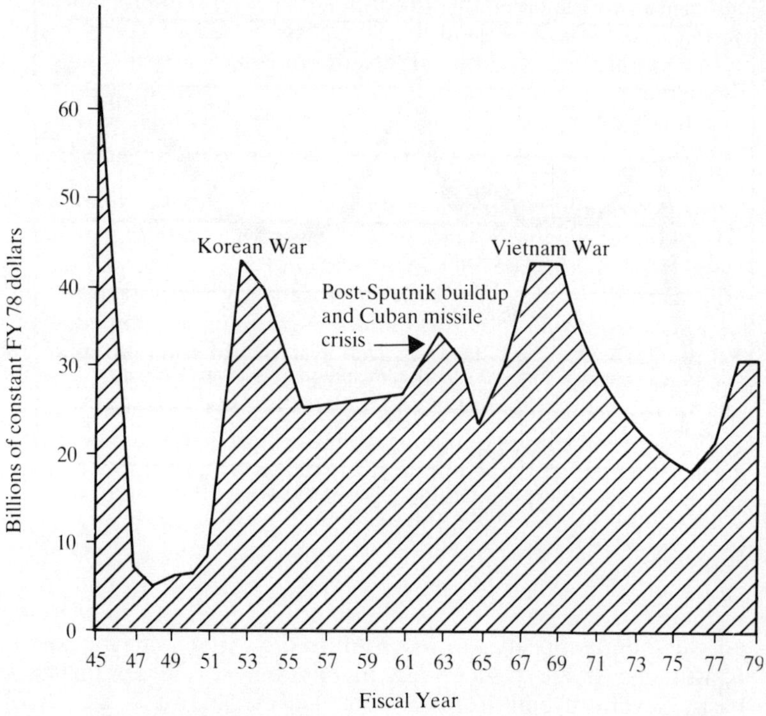

Figure 7. Evolution of Defence Procurement Since World War II.
Source. J. Gansler (1980), p.12

influence on the country's overall economic performance and rate of innovation in contrast with its major international competitors, has brought criticism. Nevertheless, one fundamental fact remains unchanged, namely, the high degree of influence wielded by the military in the postwar development of the US's R&D system.

Examining Figures 1 to 4 more closely, however, it is possible to see that this influence has not been uniform. Indeed, significant important fluctuations have occurred, generally in line with the historical pressures dominating societal development. This is particularly clear in the economic and political arenas both nationally and internationally. Figures 6 and 7 illustrate this trend by showing the variations in total defense outlays from 1950 to the early 1990s[6] on one hand, and the level of defense procurements since the Second World War to 1979 on the other. In particular, during the 1980s, military expenditure has continued to increase under the Reagan administration's policy of rearming, already raising expenditure beyond levels reached in either the Korean or the Vietnam wars.[7] Indeed, the administration's figures suggest an increase in procurement of about 240% (real terms) from 1976 to 1986.[8]

The Aftermath of the Second World War

Looking at Figures 6 and 7, it is clear that military fortunes are related to the prevalence of war and militaristic pressures. Whenever these pressures have increased, a growth in the basic resources controlled by the military has almost immediately followed. Its effect on the US's R&D system, for example, is seen in Figures 1 and 2. Indeed, if we examine the historical curve, it is clear that immediately after the war, there was a dramatic fall in the level of military expenditure from the very high levels reached during the conflict.[9] For a time it looked as if the demobilisation which followed the First World War was to be repeated with similar consequences for the social constituency of the R&D system (see Appendix). By 1945, the Federal outlays for National Defense had been almost halved and by 1947 had fallen to around 15% of their wartime peak.[10] This situation, however, was shortlived, as war soon reappeared to reconstitute the earlier wartime social constituency of the US's R&D system, this time on a more permanent basis. Several events justified what became known as the Cold War,[11] a war which, according to Freeman (1974) was largely to determine both the priorities of Government R&D policy for the 1945-70 period and the rise of the military-industrial complex.[12] The Berlin Blockade of 1948, the Chinese Revolution of 1949 and the first Soviet atomic bomb test in the same year, were all seen as a clear threat to US hegemonic interests. It was the outbreak of the 'hot' Korean War in 1950, however, which provided the most effective impetus and hence the opportunity for a surge of military spending and consequent consolidation of military influence within the social constituency.[13]

Figure 7 shows the dramatic rise of military procurements associated with the Korean War. At the same time, for the period 1950-5, national defense outlays increased from 30.9 to 76.8 billion of constant dollars, an equivalent 150% in real terms. Also during this period, total Federal outlays rose by about 35% in real terms so that, by 1955, military outlays accounted for around 60% of total Federal outlays. The impact of the military surge on the R&D system was considerable. Figure 1 shows defense and space increasing its share from 23% in 1950 to almost 50% of total R&D outlays by 1955. Most strikingly, as Figure 2 reveals, the military share of Federal funds reached 85% in 1955, suggesting a convergence of government and military interests and, more broadly, the reconstitution of the wartime social constituency of the R&D system. Corporate capital and science also played a large part in this process, benefiting in consequence.[14] Thus, as Figure 5 shows, expenditure for industrial R&D went up by 28% during 1953-5 with the Federal share for such expenditures rising from 39% to 47%.[15]

In addition, the overwhelming proportion of the financial resources accrued to the large corporations dominating the science-based industries (see Appendix). In 1957, for instance, chemical and allied products, electrical equipment and communications, and aircraft and missiles industries accounted for 66% of the total industrial R&D funds.[16] From these the electrical and aircraft industries alone had a share of 57% reflecting their prime importance for military interests. In fact, between them, these two industries used 80% of the Federal funds for industrial R&D. On the other hand, looking at the size of the companies, we find that, in 1957, those companies with more than 5000 employees accounted for 85% and 90% of the total and Federal industrial R&D funds respectively. At the same time, the first four companies with the largest R&D programmes spent 22% of the total funds and 29% of the Federal funds for industrial R&D, while the first one hundred accounted respectively for 81% and 94% of the same funds. There is little doubt, therefore, about who within the industrial sphere was benefiting most. As in the past, the corporate capital interests were evident in the reconstitution of the R&D system's wartime social constituency.

The Korean War ended in July 1953 but the social complex of power spurred by war remained strong, as is seen from military expenditure, still high long after the war had ended. Certainly, the Cold War provided the necessary long-term military pressure for this, but there was another important force, namely the idea that high government and military expenditure was a powerful means to stimulate the US economy. This led Reppy to state that the surge of military spending during the Korean War 'was only partially related to the war itself' (Reppy, 1983, p.22).

According to Chomsky,

> corporate managers who flocked to Washington to run the wartime economy learned the lesson that Germany and Japan

> had discovered without the benefit of Keynes. Government induced production of armaments, on a massive scale, can — temporarily — overcome the crisis of capitalist institutions (Chomsky, 1980, p.7).

In fact, the Second World War had rescued the US economy from the severe depressive levels of the 1930s and brought unemployment down sharply from 17.2% of the labour force to 1.2% in 1944 (Baran and Sweezy, 1975).[17] Later, however, with the general demobilisation which followed the war, reaching its peak in 1947-8, government and military expenditure decreased considerably,[18] with the result that 'the slack in the economy again appeared, bringing with it unemployment and declining economic activity' (Tirman, 1984, p.6.) It seemed clear, therefore, that high government and military expenditure was the correct policy for the country's postwar economic problems. As already seen, the Cold War and the Korean War provided the opportunity to implement such a policy, thus initiating what is generally acknowledged as a long period of unprecedented US economic prosperity which was to extend roughly until the mid 1960s.

From the early 1950s, economic and politico-military concerns reinforced each other, forming a dynamic force that provided a long-term basis on which the overlapping interests of corporate capital, government, the military and science could consolidate. Indeed, fuelled by international political and military pressures (e.g. the Sputnik challenge, the Cuban missile crisis, the Vietnam War) and their associated technological race, government, military and R&D expenditures increased markedly for more than a decade and a half.[19] It was only in the late 1960s that the effects of renewed competition from Europe and Japan,[20] the defeat in the Vietnam War and the onset of an economic crisis shook the stability of the US.

From 1950 to 1968, both government and defense total outlays rose in real terms; government outlays by approximately 240% and defense outlays by about 330% (US Dept. of Commerce, 1983). Figure 6a shows that defense outlays as a share of total government outlays reached its peak of 62.4% in 1955, falling to 45.0% in 1968. This overall decline in the share of US defense, however, was not reflected by any significant change in the military's influence on the R&D system. In my view, this decline was the result of the change in technological emphasis demanded by the Cold War and the opening of the space race with the launching of the Sputnik by the Soviet Union in 1957. In effect, considering the military implications of space technology, the Sputnik event was perceived and portrayed as a major threat to US national security.[21] Space activities were thus given high priority with a consequent transfer of resources from the Department of Defense (DoD) to the National Aeronautics and Space Administration (NASA). Figure 2 confirms this effect by illustrating the variation in the defense and space R&D expenditures within the period 1955-65. In particular, it shows that, within the Federal funds for R&D, defense and space together kept an almost constant share of 85%. Thus, what was lost by defense was gained by space in a symmetrical relation demonstrating that for a time at least the emphasis of the technological race nurtured by the Cold War had shifted spacewards.[22] On the other hand, given the low level of Federal financial support for research and development involving civilian needs — 15% for the period 1955-65 — it seems clear that this could hardly be further reduced during peacetime.

As in the past, industrial interests, particularly big corporate capital related to the military sector, benefited greatly from the enormous injection of Federal funds for military-space purposes.[23] As we shall see, this was the period which saw the birth of the semiconductor-based electronics industry, whose historical details have led Noble to suggest that the 'modern electronics industry . . . was largely a military creation' (Noble, 1984, p.7). Overall, industrial R&D expenditures rose by about 400% between 1953 and 1965, with the Federal contribution increasing steadily to reach 58% in 1960 before falling to 55% in 1965. This plainly suggests the importance of the government and the military-space sector in developing industrial R&D during those years. Corporate capital's contribution, however, should not be underestimated, since its increment was also quite remarkable: almost 300% for the period 1953-65. By the late 1960s this contribution was to become dominant as the Federal contribution began to weaken while industry's commitment remained strong.

Crisis and Loss of Confidence

By the late 1960s, the US hegemony within the world economy was showing signs of exhaustion. First, economic indicators began to show a deterioration in US performance[24] and, most importantly, a long-term economic decline vis-a-vis Japan and Europe, chiefly West Germany, began to develop. After the 1973 oil shock sent the capitalist world into its deepest recession since the 1930s,[25] this trend saw the United States' share of the Western world's production fall from 70% to less than 50% since 1950. Taking into account the share of the US transnationals abroad it would still be about 60% (Fitt, 1980).[26] In addition, the US world market share of manufactured products fell from 23% in 1960 to 18% in 1976. Meanwhile, the Japanese share increased from 6% to 15%. In terms of productivity, although in 1978 the US still had the highest rate, by the early 1980s it had been overtaken by both Germany and France with Italy and Japan close behind. Productivity in Japan's key exporting industries (steel, cars and electronics) was second to none. Even worse, the US's rate of productivity has steadily declined, dropping from an average of 2.3% for 1950-60 and 2.1% for 1960-70 to only 1.2% for 1977-83. In comparison, Japan and Germany had productivity growth averages of 9.7% and 4.6% for 1960-70, and 3.9% and 2.4% for 1977-83 respectively. Other factors show that the rate of investment was also higher in Japan and Germany than in the US. Between 1967 and 1971, it was only 7% in the States compared with 31.3% in Japan and 11.8% in Germany. In addition, the US trade balance had begun to deteriorate badly by the late 1960s. It reached $6,400 million in 1972, and rocketed to $69.4 billion in 1982, with Japan accounting for $21 billion. During 1987 the US trade deficit reached $170 billion with Japan accounting for $58 billion (*Financial Times*, 31 March 1987). US international corporations could not escape the effect of the decline. The US's share of direct foreign investment dropped from 53.8% in 1967 to 47.5% in 1976. At the same time, Japan and West Germany increased their share from 1.4 to 6.7 percent and from 2.8 to 6.9 percent respectively. By all appearances, as Norman put it, 'the United States has seen its dominant position greatly eroded' (Norman, 1981, p.49).

There are various reasons for the relative decline of the US economic hegemony. Among the most important are the structural factors whose influence was seen in a significant fall in profit rates in the most advanced industrial nations, even before the oil shock of 1973 [Ernst (1983), Kaplinsky (1984)]. These factors include rising production costs due to increases in wages; a saturation of the market for major consumer goods (e.g. cars and televisions) and a reduction in the demand for products made by industries which had been important during the postwar expansionary period (e.g. chemical and metal industries); also, new patterns of international competition expressed by Japan and Europe in their growing challenge to US dominance within its own market as well as in the world market. Less important, but still relevant, was the emergence of the Third World's newly industrialising countries (NICs) challenging the markets for products of more traditional industries, e.g. textile, garments, shoes and leather (Molina, 1987).

Perhaps most significant, however, is that the US leadership of the fifties and sixties was simply the result of the temporary elimination of competition caused by the war. In Norman's words 'The US technological dominance was somewhat artificial, and it was bound to erode following the successful rehabilitation of the war-torn economies of Europe and Japan' (Norman, 1978, p.14). The other relevant event was the Vietnam War. As seen from Figures 6 and 7, this war brought an increase in the level of defense expenditures as large as that of the Korean War. This time, however, the conditions were different, and with Japan and Europe greatly recovered from the effects of the Second World War, its impact was adverse,[27] not least because the adventure turned into sour defeat. Thus, while the US military was stretched worldwide, Japan and Europe, chiefly West Germany, relieved of a large burden of military expenditures by the presence of US troops, began to assert themselves on the economic front (Vigier, 1980).

As the war ended, the protest movement it raised[28] and the inevitable demoralisation and questioning which followed defeat had a powerful impact on the R&D system and the social constituency which had controlled its development thus far.[29] This was to be a new period marked by loss of confidence in the power complex and the emergence of 'a new set of demands and values' which effectively constituted 'a reaction against the forces that had shaped technological development during the postwar years' (Norman, 1981, p.56). It was, in the words of Dickson, a

> critique of the social consequences of unfettered technological development, ranging from the environmental damage caused by the side-effects of modern science-based production processes to the use of sophisticated electronics in the war in Vietnam (Dickson, 1984, p.30).[30]

Increased concern for social responsibility in the development of science and technology was soon reflected in the control and allocation of the basic resources of the R&D system and, ultimately, in the position and influence of its dominant social constituency. Figure 1 reveals that the first impact was a stagnation of R&D expenditure from the late 1960s to the mid 1970s. As a result, R&D expenditure as a percentage of Gross National Product (GNP) dropped from a high of 2.9% in the mid 1960s to a low of 2.3% in the mid 1970s. In parallel, from Figure 1 we also see that the total Federal contribution as a percentage of total R&D outlays began a decline which by 1980 was still continuing. An important part of this drop is accounted for by the decline in the

Federal defense and space expenditure which fell from its peak of 85% of the total Federal funding for R&D in the early 1960s to a low of 60% by 1980 (see Figure 2). The other factor was that non-Federal outlays, particularly industry's, increased during the same period, thus reinforcing the relative drop in the share of the Federal contribution. Figure 4 shows how, between 1960 and 1975, industry's funding of R&D rose while government's fell. It also reveals that this rise of industry as a source of funds was roughly enough to prevent a fall in overall R&D performance. For this reason, over the decade (1965-75), R&D expenditure in the US remained mainly stagnant.

During the 1965-75 period therefore we see a distinct alteration within the social complex of power dominating the development of US's R&D system. First, as war and international politics lost impetus and the interests — 'new set of demands and values' — of other social forces gained strength, for the first time since the war an effective opposition challenged the interests underlying the postwar power complex. As a result, the relative weight of the military-space constituent vis-a-vis other forces, and within the government, weakened considerably. Admittedly, as can be seen in Figure 2, the Defense share of Federal funds for R&D remained around 50% from 1965 to 1980, while it was the space budget which suffered the brunt of the cuts, mainly, it is argued, because the Apollo programme and the race to the moon had come to an end in the late 1960s. The point, however, is that, in the sociopolitical conditions of the late 1960s, even if this was correct and the space budget had to drop 'naturally', the percentage lost by space could not go back to defense where it had originally come from. The pressures from the 'new set of demands and values' called for more government involvement and support for R&D with direct social relevance. Thus, as illustrated by Figure 2, the percentage of total Federal funds devoted to civilian purposes rose from 16% in 1965 to 40% in 1980.[31]

But the military's reduced control of the R&D system's resources was seen not only in the system's altering social goals, but also in the ascendancy of industry's and hence, corporate capital's relative weight within such a system, to the point that, by 1970, industry's share of funds for industrial R&D had reached 57% from 45% only five years before (Figure 5). Most dramatic, however, a decade later, in 1980, industry's share of funds for all R & D in the US overtook that of the Federal government (see Figure 4). Since this trend has continued into the 1980s, it means that, overall, the control of the basic resources of the US's R & D system has shifted in favour of corporate capital, particularly that from high technology industries. Figures 3 and 4 confirm that, in 1982, industry provided almost 51% of all the financial resources devoted to R & D while its laboratories spent more than 73% of all these resources. In addition, in line with the historical pattern, there was enormous concentration of these resources in large science-based industries. Corporate capital has thus become dominant in the development of the research and development system. While industry had always carried out a large part of the R&D programme, now it has also become the dominant source of funds. In the next chapter, we shall find that this has important implications for R&D in the electronics industry, supporting my contention that R&D activity is shaped by those interests who exercise control of its basic resources.

The Resurgence of the Power Complex

By the second half of the 1970s, the 'new set of values and demands' which had emerged in the late 1960s had lost momentum. They were increasingly pushed into the background by the spiralling economic crisis following the oil shocks of the 1970s, and by the realisation that the US economy was in decline. International competition rather than unchallenged US hegemony became the pattern for the future, with the result that demands for greater social responsibility and accountability in the development of science and technology were superseded. Such demands could not provide an impetus similar to war or international competition – either politico-military or economic. This period of socially-aware demands was a time of little growth for the R&D system. As soon as the spectre of the Vietnam defeat began to fade, replaced by alarm at the economic crisis, war and international competition began to reassert themselves and, with them, the same dominant social interests. As Dickson has put it,

> The postwar period saw decision-making over the allocation of funds for science largely dominated by scientific, corporate and military elites . . . In the late 1960s and early 1970s, the domination of these three groups was challenged . . . Now, waving the banner of social efficiency and international competition, and with the direct encouragement of Washington, these three elites are re-establishing their alliance . . . Industrial leaders argue that only scientific and technological supremacy over the rest of the world will allow the country to prosper economically. Military leaders claim that only a rapidly increasing military research budget, feeding directly

> into ever more sophisticated weapons of mass destruction, will ensure a stable peace. Politicians have picked up and faithfully amplified both refrains (Dickson, 1984, pp. 18 and 3).

By the early 1980s, there was clear evidence that the convergence of social interests which has dominated postwar developments in the US was back in strength. A combination of military and economic pressures which may have contradictory effects on the development of the US's R&D system, had provided the impetus for its resurgence.

Since 1980 there has been considerable rearming in the name of the US and her NATO allies' security. It has been argued that during the 1970s the US military capability dangerously deteriorated relative to the Soviet Union and that this was destabilising for a world peace based on the balance of deterrence. Fears about Soviet aggression and expansionism were fuelled by the USSR's invasion of Afghanistan in December 1979 and the military coup in Poland in 1982. As a result, the military reasserted itself, seen most clearly in the increase of basic resources under its control. Military expenditure has increased markedly since 1980 (see Figures 6 and 6a) and is heading to surpass even the high levels of the Vietnam period. Indeed, in the words of Cain and Adam,

> By fiscal 1991, according to projections by the Office of Management and Budget (OMB), military outlays would be higher in constant dollars than in any year, in peacetime or wartime, since the end of World War II (Cain and Adam, 1987, p.50).

From a different perspective, according to SIPRI analysts, given the projections of the US administration 'by 1988 the US military expenditure would have doubled its volume in the course of one decade' (SIPRI, 1985, p.270).[32] Not surprisingly, in relation to the R&D system the military influence is again increasing, particularly as a proportion of Federal expenditures. Figure 1 shows an increase from 29% to 36% in the defense/space share of total R&D expenditures for the years 1980-5 and a similar fall in civilian pursuits, whereas Figure 2 shows an estimated increase of 23% in the defense's share of Federal funds for R&D for the period 1980-5, with a drop of 15% and 7% for civilian and space pursuits respectively. In this context, the invasion of Grenada in 1983 brought a renewed sense of global power, a growing feature of the more militaristic mood of the eighties.

In economic terms, efforts have been concentrated on revitalising the US technological and industrial base as a means of strengthening the nation's competitive muscle in world markets and, ultimately, to counteract the trend towards a relative decline, internationally, in the US economic performance. In this respect, we have already seen the details and reasons for the decline of the US's postwar hegemony. In the eighties, however, it is clear that the search for prosperity similar to that of the fifties and sixties is taking place in a different arena. The presence of strong industrial competitors in the world markets has virtually eliminated the unchallenged freedom of that halcyon period, and it has been left to US corporate capital to respond to the technological challenge imposed by international competition and, more broadly, by the process of capital accumulation as a whole.[33] Figures 3 and 4 show that research and development in industry have picked up and have been doing so since the late 1970s, after the 1965-75 stagnation.[34] In terms of performance, industry has kept its share of the increased total funds for R&D to around 75%, while, as a source of funds for industrial R&D, its share has remained around 68% (see Figure 5). As expected, large corporate capital has taken the lion's share of the renewed drive for R&D with the first hundred companies (in terms of size of their R&D programme) accounting for 80% of the total funds for industrial R&D in 1982. Of these, the first four alone accounted for 20%. With regard to the Federal funds for industrial R&D, the concentration is even greater with the first hundred companies accounting for a share of 95% and the first four companies 20% of these funds in 1982 (NSF, 1984). And to complete the historical pattern of the US's R&D system, most of the funds have been spent in science-based industries. Again, in 1982, the chemical, machinery (including computers), electrical, aircraft and missiles industries accounted for almost 70% of the total funds and almost 85% of the Federal funds for industrial R&D. The aircraft and missiles industries alone accounted for 44% and 77% of the total funds and Federal funds respectively. And, reflecting the military drive, their share reached 53% in 1982 (NSF, 1984).

It would seem that the postwar social complex of power, having re-encountered favourable economic and politico-military pressures, has again achieved a convergence of overriding interests which, as in the fifties, might lead to simultaneous economic and military strength. The government is providing large financial resources, and the military is channelling most of these towards the corporations and scientific centres in their demand for sophisticated and expensive weapons systems. At the same time, industry is benefiting the other constituents in its effort to increase productivity and enhance the competitive strength of the United States economy.[35] It would seem, therefore, that a sort of virtuous circle of self-reinforcing interests, driven by military and competitive pressures, has been found, and which may restore the United States to its leading position of the 1950s.

Contradictory Trends

The situation, however, is not so straightforward. The simultaneous presence of strong military and competitive pressures tends to have contradictory effects upon the development of the R&D system. After the Second World War, there was little international competition. Thus, the military and international political situation remained paramount in shaping the development of the system. This was also the time when the military's influence was at its peak. Today, in contrast, not only does corporate capital carry the greatest weight within the social constituency but, most importantly, the presence of Japan and Europe and even emerging Third World countries in the world markets, is making it increasingly difficult to accommodate the military's enormous demands upon the R&D system without sacrificing resources from the commercial battle. The development of the United States' military and commerce depends on R&D resources. For both sectors, high-technology is the key to success. The military's demand is for smarter and more deadly weapons, thus counteracting any advantage held by the Warsaw Pact forces in numerical terms. The US has a definitive lead over the Soviet Union in the high-technology fields relevant to military applications[36] and the government and the military wish this to expand and, above all, be translated into military advantage [Perry and Roberts (1982), Marshall (1981)]. On the other hand, the commercial claim on the R&D system is rooted in the US economic performance, and recent figures clearly suggest that the most dynamic industrial sectors are those related to high-technology. Figure 8 illustrates, for example, the difference in productivity between the US semiconductor industry and the manufacturing sector, and the US economy as a whole. More generally, as Staat has put it

> When high- and low-technology industries are compared, high technology firms have productivity rates twice as high, real growth rates three times as great, one sixth of the annual price increases, and nine times the employment growth. The same kind of favourable ratio prevails in terms of international trade. The trade balance for R&D-intensive

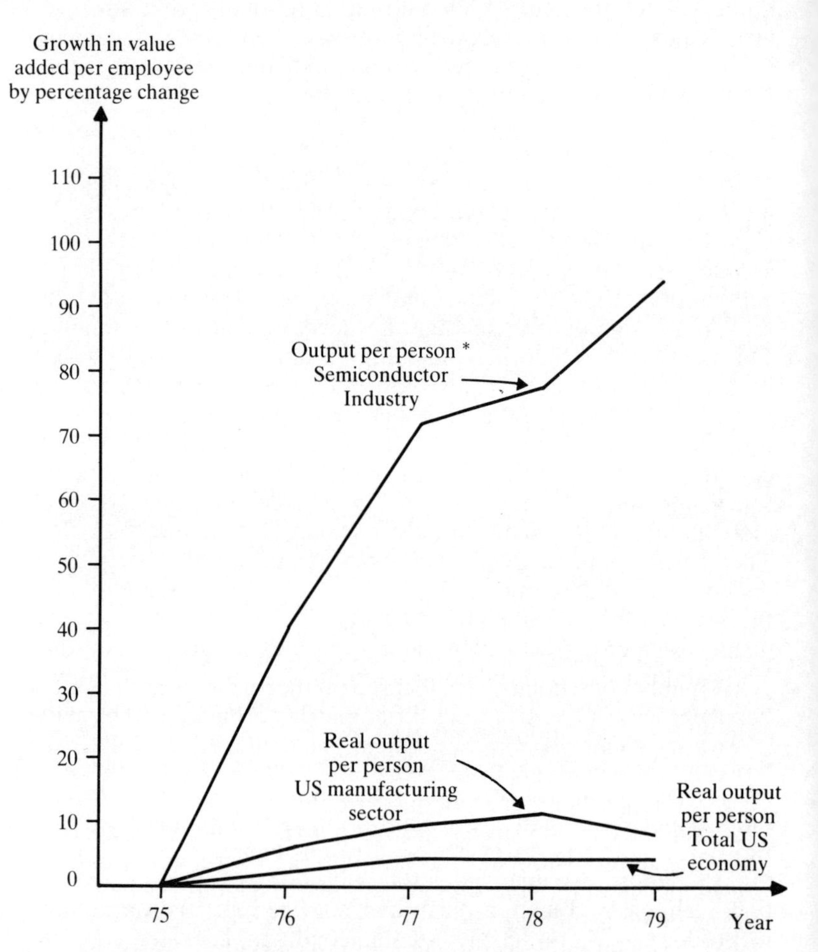

Figure 8. Comparison of US Productivity Trends for the Semiconductor Industry, Manufacturing Sector and Total Economy

* Due to the lack of real value estimates for the Semiconductor Industry, the comparison uses nominal or current dollar estimates for this industry and real value estimates for the total economy and manufacturing sector. The effect is to yield a conservative estimate for the growth in productivity in the semiconductor industry, given that its traditionally declining unit prices means that adjusting for price changes would produce an estimate for real output substantially above the level of nominal output

Source. Based on figures given in W. Finan (1981), p.25

> manufactured products has been generally rising through the period 1960-1976, and is now more than $28 billon. The trade balance for non-R&D-intensive products is down from a break-even level in 1960 to a 16 billion deficit. (Staat, 1979, p.136.)[37]

Both economic and military objectives have a legitimate claim to the resources of the R&D system. However, since the products, associated demands and end-users are different for both cases,[38] they seem to lead not to a unified and coherent path of development, even when financial resources are being poured into similar technological areas, but to separate paths where the main interconnection is the eventual 'spin-offs' from one side to the other. From past experience, there is little doubt that, under the current military drive, spin-offs will occur. The big question, however is: given both the large demands of the military upon the US's R&D system and the presence of strong commercial competitors whose military concerns are considerably less than those of the United States,[39] will the spin-offs mechanism coupled with industry's own commercial R&D be sufficient to enable the US economy effectively to stave off the growing technological challenge in the international markets? Undoubtedly, it is too early to say,[40] but the evidence suggests that this may be a rather difficult undertaking.

The Case Against the Military Burden

A number of scholars have focused on the impact of high military expenditures on the country's economic performance and have put forward a technoeconomic critique of the militaristic approach and its associated spin-offs theory.

Most typically, the economic critique suggests a correlation between the United States' relative economic decline in comparison with Japan and Europe, and the influence of the military in the development of the country's R&D system and industrial base. This is argued on two levels.

At the quantitative level, comparisons are made between the US's military burden and economic performance and those of Japan and Europe, notably West Germany. This shows a negative correlation between high military control of basic resources and economic performance. For instance, Japan, with the smallest percentage of Gross National Product (GNP) devoted to military purposes has also had the highest growth rate among developed countries, whereas the United States, with the highest percentage of GNP for military purposes has achieved only moderate growth rates [OECD (1971), Niosi and Faucher (1985)]. More specifically, it is argued that due to the heavy US military burden, Japan invests far more of her capital resources in constantly renewing her civilian base than the US does — hence Japan's success in international competition.

In 1979, for example, the ratio of military to civilian use of capital resources in the States was 33 to 100, while in West Germany it was 20 to 100 and in Japan 3.7 to 100. Following the US military build-up of the 1980s, the United States' position has become even worse with some analysts estimating a sharp increase in the ratio of military to civilian use of capital (Melman, 1986). Already in 1982, according to Dumas (1986), the Defense Department's net investment in industrial equipment was nearly 38% of the average annual net investment by all US manufacturers combined. The same author also points out that, in 1983, the total reported value of all physical capital owned by the military, excluding land-holdings, was about $470 billion. By comparison, the value of the combined net stock of all equipment, structures, and inventories owned by all US manufacturing establishments was about $1.012 trillion. Thus, in 1983, the book value of the physical capital preempted by the government-owned part of the military sector was 46% as large as the value of the total capital stock of all US manufacturing establishments combined. Since this figure does not include the value of physical capital privately owned by military industrial manufacturers, in practice the military control of the US basic resources turns out to be even greater.

Similarly, the military-burden critique shows that the US and the UK, heading the list in military-related R&D expenditures among capitalist powers, are simultaneously at the bottom in terms of productivity growth [De Grasse (1984), Barnaby (1981)]. In the period 1961-75, Japan and West Germany spent on defense and space 4% and 20% of their respective overall government research and development budgets. In the United States the average for the same period was 70% (Dumas, 1984). By conservative estimates, this means that, in the US, the military actually control about 30% of the R & D effort administered by universities and colleges, 30% of industrial R&D activity and the talents of a third or more of the nation's pool of scientists and engineers [Dumas (1984), Melman (1986)]. Not surprisingly, it was found that, in 1982 for instance, the Japanese civilian R&D effort received nearly 56% more newly graduated engineers than did the United States, a country with almost double the population of Japan. Currently Japan's ratio of non-defense R&D expenditure to gross domestic product is 2.6%, the highest in the world in spite of the fact that, with $57 billion for R&D in 1986, they spend less than the United States (*Financial*

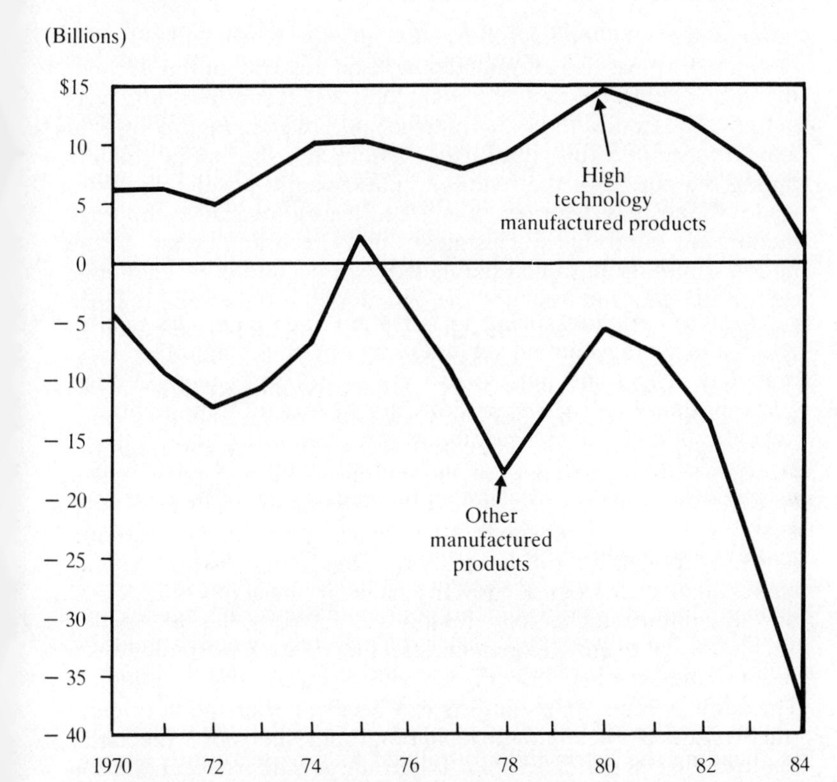

Figure 9. US Trade Balance in High-Technology and Other Manufactured Product Groups. Constant 1972 dollars
Source. National Science Board (1985). *Science Indicators. The 1985 Report,* US Govt. Printing Office, Wash., DC

Times, 15 July 1987). Finally, it is also argued that the rate of innovation has also shifted against the United States, chiefly in favour of Japan and West Germany as shown by the decline of the US's share of patenting both at home and abroad. [41] The conclusion is that, in the US, the capital and technological brain-drain generated by the military sector has gradually led to the relative technological decline of the civilian sector and, hence, to a decline in the country's relative economic performance. This conclusion is not shared by other analysts (see Adams and Gold, 1986) and, undoubtedly, the process of relative US decline is a complex affair whose solution involves more than a simple reduction of the military burden. But, for Niosi and Faucher (1985), who have also examined other explanations for this decline, the military-burden argument is the one that seems 'to fit the facts better'. The trends in the US trade balance given in Figure 9 would appear to support this view further. There, it is shown that roughly during the period of the renewed military build-up, the country's trade balance has deteriorated markedly even for high-technology manufactured products. This means that the strengthening of the nation's military has been paralleled by a decline in its economic competitiveness. In fact, for the first time the US's trade balance in high technology went into the red with a $2.6 billion deficit in 1986 (*Financial Times*, 20 May 1986).

At the qualitative level, the technoeconomic critique raises various points which, ultimately, would explain the US's poor economic performance. The most crucial of these concerns the growing inappropriateness of the military technology for civilian applications and, thus, the difficulties affecting the transfer of technology from one sector to the other which, given the high military demands upon its basic resources, would lead to the distortion of the US's civilian technological base. In particular, the inappropriateness of the military technology relates to the so-called process of 'technological overdevelopment' affecting military technology as a result of the exacting performance criteria imposed by the military. This view accepts that spin-offs can be created at the very early stage of development of a technology but the more the latter matures the less possible spin-offs become as military specialisation takes it farther and farther away from commercial applications. In addition, this view suggests that overdevelopment is associated with economic concentration in the military industry[42] and hence, tends to inhibit innovation and efficiency (Tirman, 1984). Spin-offs are also affected by the lack of communication and technology transfers from the military to the commercial sector due to basic differences between the commercial and military environments: there is little price sensitivity in the military arena while cost is the central concern of the commercial customer (Melman, 1986). In addition, there is conflict between the military custom-made requirements and the civilian need for standardisation (De Grasse, 1984).[43] To make matters worse, secrecy, concerns for national security and lack of incentive to transfer technology from defense to the civilian sector have all diminished the possibility of spin-offs (Tirman, 1984). Ultimately, all these factors are at the core of the contention that the high military control of the US basic resources for science and technology actually leads to a distortion of the civilian technical base. Indeed, if we consider both the arguments outlined, it seems clear that the drain of all these resources away from the civilian and into a military realm is bound to affect the development of the civilian technical base in comparison with strong competitors whose military burden is substantially smaller. As Botkin *et al.* (1982) have expressed it in the context of the Reagan administration,

> International competition and Reaganomics have caught the high tech industry — and with it the American economy — in a squeeze play. While Japan challenges our technological leadership, burgeoning defense programs soak up engineering skills critical to continued American innovation (Botkin *et al.* p.2).[44]

Undoubtedly, the above arguments will dominate the development of the United States' R&D system for years to come. The final word, however, belongs to history. For this study the most relevant aspect of the postwar process is that it has shaped microtechnology development in the United States. In the following chapter we shall see that, step by step, the development of microtechnology has been inextricably related to the social forces and issues we have encountered above.

THREE

THE SOCIAL SHAPING OF UNITED STATES' MICROTECHNOLOGY FROM THE SECOND WORLD WAR TO THE MID 1970S

As indicated earlier, the process of social shaping of microtechnology represents a particular case of the process we have discussed in relation to the research and development system. This means that the issues, trends and evidence already dealt with in the previous chapter will help towards understanding the process involving microtechnology. The argument covering microtechnology follows the same contention as for the case of the R&D system: namely, that the post-Second World War capital-government-military-science complex of social interests has effectively been the dominant social constituency behind the development of US microtechnology. As a result, this development has largely mirrored the convergence of interests, tensions and relative influence of the dominant constituents under changing circumstances.

The postwar history of microtechnology reveals it as a focus for the convergence of interests of the postwar social complex. This convergence of interests achieved clearest expression during microtechnology's infancy, following the invention of the transistor and the subsequent search for miniaturisation, leading to the integrated circuit (IC). This period witnessed the early convergence of electronic systems on the basis of semiconductor techniques. It also saw the consolidation of both the social constituency of microtechnology and, more specifically, the modern electronic industry, geared to producing microtechnology for a commercial outlet. Below I outline the influence of historical events on the role of the social constituents in shaping microtechnology and the electronics industry.

The Legacy of the Second World War

The dominant influence of the military in the early social constituency of microtechnology has been the theme of such authors as Noble (1984). Admittedly, it was not the military alone, but the entire social complex of power which actually created the modern electronics industry. Nevertheless, it was the military who played a leading role in shaping the technology's development in a process which started with the Second World War.

The previous chapter showed how this war stimulated the wartime social constituency of the United States' R&D system through the scientific and technological efforts demanded by large-scale military enterprise. One of the main beneficiaries of these efforts was electronics technology, major programmes concerning components and systems being carried out in many areas. As *Electronics* put it, 'the exigencies of World War II transformed electronic technology . . . Out of the conflagration were born computers, miniaturization, radar, loran, and guided missiles' (*Electronics,* 1980, p.151). The radar programme in particular was a huge undertaking estimated to have cost $2.5 billion in comparison with $2 billion for the A-bomb Manhattan project. The radar project more than any other during the war epitomised the emergence of the wartime social constituency of microtechnology. Stimulated by the war and the huge financial resources made available by the government and the military, major R&D centres such as Bell Labs and the Massachusetts Institute of Technology Radio Laboratory as well as industry became heavily involved in radar work.

Thus, at one time, one-fifth of all the United States' physicists worked on radar development at the Radiation Laboratory, while such large corporate establishments as the Bell System and Westinghouse advanced its development and implemented its production [*Electronics* (1980), Brooks (1976), Noble (1984)].[1] In radar, therefore, government, the military, science and corporate capital converged, drawn together by war into an interaction which not only satisfied the common interest of winning the war but also the particular interests of each party. Industry, for instance, received financial resources for R&D work which ultimately would strengthen its technical capability, while the manufacturing contracts provided a secure market for its production.[2] Likewise, science saw its financial resources and facilities expand with the creation of such new centres as the MIT's Radiation Laboratory and the Radio Research Laboratory at Harvard University. Finally, government and the military power saw the country's economy and military power thriving as they carried out the US's role in political, military and economic affairs, at home and abroad.

The expansion of the electronics industry during the war was decisive, establishing its position as a permanent factor in the US economy and technical base. According to *Electronics* (1980), it expanded

> to more than 12 times its monthly prewar net factory filling value on end equipments and from over 2 to 20 or 30 times on the major components. The expansion of labor employment was from a prewar peak of approximately 110,000 workers to a war peak of 560,000 (*Electronics*, p.158)

Most important, however, were the qualitative advances in the technology which laid the foundations for the present-day electronics industry. During the war, important advances took place in components, miniaturisation techniques, computers and control systems. The performance of electronic valves was improved and, most significantly for the subsequent invention of the transistor, radar work meant the heavy use and improvement of crystal rectifiers — notably Germanium — as mixers, video detectors and dc restorers. In turn, the development of the proximity fuse led to the search for miniaturisation techniques which prepared the way for the integrated-circuit era.[3] Finally, computers and control systems also underwent crucial advances. The first digital computer, ENIAC, for instance, was the result of a military-science project aimed at solving the difficulties of ballistic calculations.[4] During the war, various relay-operated digital computers were designed for the Armed Forces, in which C. Stibitz of Bell Labs played a major part (Goldstine, 1972). Lastly, in 1944 the Harvard-IBM ASCC or Mark 1, although mainly a corporate capital-science effort (Pugh, 1984), also entered full-scale operation, tackling ballistic problems for the Navy (Fleck, 1973). With control systems, efforts to develop weapon-fire control systems rapidly advanced the technology of closed-loop or servo-systems.[5]

Industry and universities were the centre of these efforts with Bell Labs and the Servomechanism Laboratory at MIT taking the lead. Thus, in 1943 Bell Labs produced for the Army 'the first practical antiaircraft system that was essentially automatically controlled' (Brooks, 1976, p.212), while in 1944, the MIT's Servomechanism Laboratory, alongside its work on automatic fire control, also began to build a simulator for multi-engine aircraft at the US Navy's request (Fleck, 1973).

The impact of the Second World War upon electronics was momentous.[6] As one commentator says, it 'had generated a wide range of components and devices, greater understanding of electronic technology, and an army of electronics enthusiasts' (Noble, 1984, p.47). Most crucially, it had also generated a constituency of social interests, which had come to control and shape the development of electronics technology, largely towards war and military needs. Finally, for the first time on a major scale the spillover effect from military into civilian technology had become an accepted way to develop electronics. Only the end of the war and a decisive, long-term reduction of military pressures might have altered this situation, but this was not to happen in the next two decades.

The Government-Military-led Development of the Postwar Period: 1945-mid 1960s

As with the United States' research and development system, the end of the war brought a sharp decline of the military burden on the country's industrial base. This was reflected in the development of the electronics industry, where the government and military constituents were rapidly displaced from their dominant, directive role. This is illustrated by Figure 10 which shows that, by 1950, the government's share of the market had declined to less than one-fourth of the total sales of the US electronics industry. As two analysts have commented, 'The government had naturally been the biggest purchaser during the war, but was rapidly toppled by the post-war consumer spending spree' (Braun and MacDonald, 1978, p.78).

This trend did not last. The Cold War and the Korean War were soon to restrengthen the wartime social complex of power. Figure 10 shows the impact of this process on the development of the US electronics industry. From 1952, it is possible to observe that, once again, the government's share of the market had become dominant. This situation continued for many years. In the process, the technologies and the industries behind today's microrevolution were shaped by the interests of a social constituency most heavily influenced by the military, as was the case for the R&D system. This fact is strikingly illustrated by Figure 11 where a broad range of advances in electronics systems, between 1952 and 1966, is shown to have originated from an equally broad range of military projects.

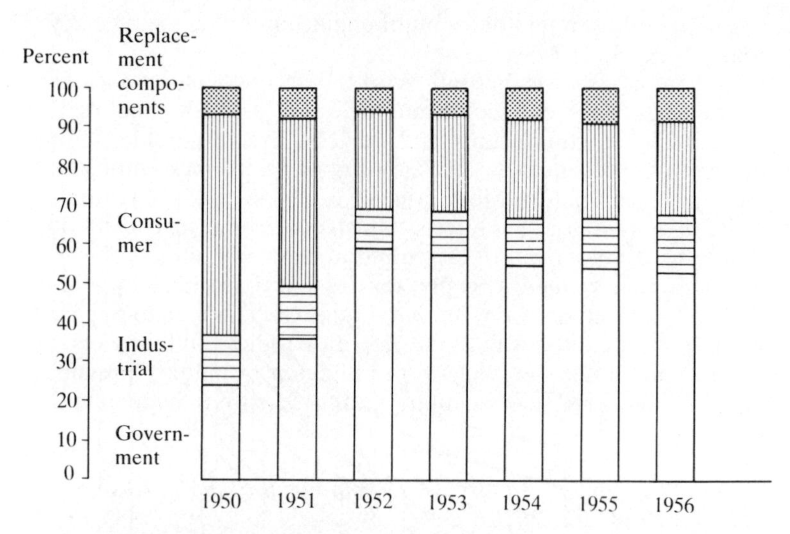

Figure 10. Percentage Distribution of US Electronics Industry Sales by End-Use (1950-56)

Source. Based on figures given in E. Braun and S. MacDonald (1978), p.79

The influence of the government-military constituents was wide-ranging, affecting all areas of electronics. Nowhere was this influence more decisive than on the development of the emerging technologies and industries of computers, industrial control systems and, above all, semiconductors, the technical base for the convergence of electronics systems. In this sense the telecommunications technology and industry stand apart, since their well-established development began in the late nineteenth century. Telecommunications had none of the open-endedness of emerging industries, in fact corporate capital had fairly tight control of the field, particularly with the Bell System's monopoly of the telephone industry. The Bell System, for instance, was described in 1939 as 'the largest aggregation of capital ever controlled by a single company in the history of private business enterprise' (Coon, 1939, p.2). At the time, it was a five-billion-dollar business with nearly 650,000 stockholders and annual purchases running into hundreds of millions of dollars (*ibid.*).

A corporate organisation of this size, exercising a near-monopoly of the United States telephone system, was clearly as powerful as government and military interests. Bell had a well-established technological base of its own and, although it benefited greatly

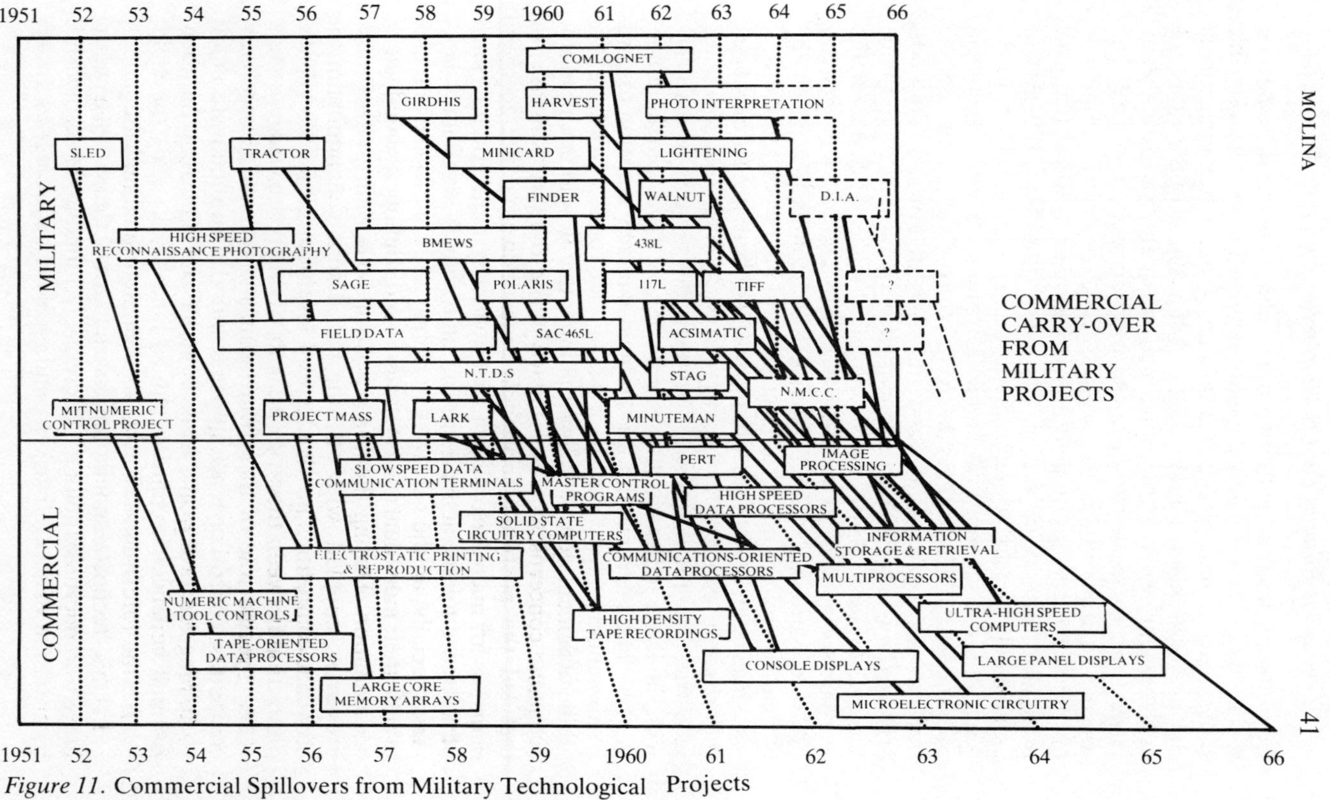

Figure 11. Commercial Spillovers from Military Technological Projects

Source. J. Diebold (1964), p. 37

from its connection with the other members of the wartime social complex, it did not depend on them for its development. Bell's main technological commitment was to telecommunications. Because of its vast resources, it could keep control of the industry's development, while at the same time incorporating the advances made from its participation in the wartime social complex. The war had seen the Bell Labs 'forged . . . into probably the greatest invention factory in any field of science that the world had ever seen' (Brooks, 1976, p.216), and after the War, the system immediately began to reap the benefits from its R&D work. Thus, the first commercial mobile telephone service was put in operation in 1946; in the same year, coaxial cable was used to transmit television signals; and, in 1947, microwave radio was first used for long-distance telephony (*ibid.*). With the Cold War, the Bell System again became an important participant of the wartime social complex (see Brock, 1981), and although military pressures certainly shaped important technological work in the System, the effect upon the overall development of the industry was small compared to the extent to which these pressures shaped the nascent computer, industrial control and semiconductor industries. In the following discussion my concern is basically with the latter industries whose technological development during the fifties and early sixties was largely shaped by the military and government.

The argument goes as follows. Stimulated by war and politico-economic concerns, it was the government-military constituents that led the postwar social complex of power. Through their demands for microtechnology they provided not only a clear frame of reference for the activities of corporate capital and science but, most crucially, all the necessary resources and conditions for these to fulfil their individual needs for capital and scientific-technological accumulation. During this time, political, economic and military pressures, coupled with the virtual absence of international economic competition provided a fertile environment for the smooth interaction between these constituents of microtechnology.

The specific role of the government and the military towards computers, control systems and semiconductors was multi-faceted, combining technological and economic factors. In addition, they fostered beneficial conditions, for instance by keeping the Cold War at the forefront of the country's international politics.

In this technoeconomic approach, three factors had most impact on microtechnology.[7] These are: a) funding of R&D work; b) provision of secure markets; and c) broad and sometimes very precise specification of technical demands. In varying degrees, these three aspects accounted for much of the 1950s and early 1960s development of the computer, industrial control and semiconductor technologies and industries.

a) Industrial Control

In the field of industrial control, the origin of numerical control (NC) — the automatic control of machine tools by means of numerical data stored on punched tape or cards — and the growth of an NC machine tools industry can be attributed to a military-led social constituency exploiting the technical advances of the Second World War. In effect, in 1949, the wartime experience in gunfire control systems brought Massachusetts Institute of Technology's Servomechanism Laboratory to the centre of an effort to develop an NC milling machine for the US Air Force (USAF). The idea came from Parsons Co., Michigan, informed by the USAF's technical specifications for a wing panel for a new combat fighter [Braverman (1974), Noble (1984)].[8] The machine was developed at MIT and together with the USAF they undertook to promote NC technology among the established machine tool industry (NAE/NRS, 1983). Machine tool builders, however, reacted cautiously to the prospect of risking their own capital in the new technology. Only one company, Giddings and Lewis, responded (Noble, 1979).[9] In the aftermath of the Korean War, however, all this was to change, as Layton (1972) describes,

> in 1956, following the Korean War, the US Federal Government took the decision to build up a stock of advanced machine-tool capacity, both as a reservoir for potential armaments production, and as a means of stimulating advanced technology. The US Airforce approached the machine tool and electronic manufacturers, first with development contracts for numerically controlled tools and then with orders for batches of 50 — in all worth $30 million. Bendix, General Electric and Thompson Ramo Wooldridge (later Bunker-Ramo) obtained contracts for numerical controls and Cincinatti for tools. Software was further developed at MIT under government contract (Layton, pp.171-2).

Thus, the USAF not only subsidised NC research but single-handedly created a market for it (Schlesinger, 1984). As a result, the military-led social constituency went into full operation so that, by 1949, the commercial market for NC machine tools took off, a spillover from the military market. By the late 1960s, between 15%

and 20% of the value of machine tools installed each year in the US was numerically controlled (*ibid.*). In this, the military's role had been paramount. They had provided the technical problem, the funds for research and development work, the secure market which enabled the convergence of interests of all the social forces controlling NC technology's resources.[10] In industrial terms, there was little doubt who the main beneficiary was. As Schlesinger said, 'the benefits of this effort redounded almost entirely to the largest companies in the computer, machine tool, and control systems fields, as well as to a limited number of big aircraft firms' (Schlesinger, p.184). Thus, within industry, it was large corporate capital which played the leading part in the social constituency of the NC technology.

b) Computer Systems

The case of computers follows a similar pattern to that of industrial control, with the government-capital-military-science concert of interests plainly interacting in the R&D, production and demand for these systems. Building on wartime advances, it was the government and military who again pioneered the drive for computers by providing the resources and conditions for their development. Indeed, as Sharpe states,

> Until 1951, the computer industry was essentially non-commercial: each machine was one of a kind, and support came primarily from universities and government. In fact, it can be plausibly argued that without government (and particularly military) backing, there might be no computer industry today (Sharpe, 1969, p.186).

It is possible to say that the whole of the first and an important part of the second generation of electronic computers, until the early 1960s, were largely the result of government and military involvement.[11] On the one hand, their demands provided the focus around which most computing development revolved, and, on the other, their support enabled and stimulated the participation of scientific and corporation interests, not only through the necessary funds for R&D but also through 'a ready market due to federal procurement of computers' (Soma, 1975, p.3). In this way, they spearheaded the formation of a social constituency which included, with themselves, the considerable body of computing scientists which had developed during the war,[12] and big corporate capital, chiefly from those industrial areas most likely to be affected by computer development.[13] In terms of computer research and development, the military and government role is described by Schnee

> From 1945 to 1955, the US Army, Navy and Air Force, the Atomic Energy Commission, and the National Bureau of Standards all placed major contracts for the development of improved computers with universities and with the firms who began design and manufacture — especially Remington Rand (Univac) and, later, IBM. As a result, rapid progress was made in solving some of the problems of logic design, memory storage systems and programming techniques (Schnee, 1978, p.16).

This process manifested itself in a variety of interactions between the social constituents, all of which helped to shape the modern computer industry. A major form of interaction was based on large-scale computer projects for clearly specified military purposes. The outstanding example was the Whirlwind-SAGE (Semiautomatic Ground Environment)[14] which, at different stages, involved the Navy, USAF, MIT's Lincoln Laboratory (previously Servomechanism Laboratory), IBM, Burroughs, the Rand Corporation and Western Electric (Bell System) [Katz and Phillips (1982), Soma (1975), Dinneen and Frick (1977)]. It began operations in 1958 and enormously advanced computer technology, particularly regarding real-time operations. It is also acknowledged that IBM owes much of its current position as the giant of the industry to its prominent participation in the SAGE project (Katz and Phillips, 1982).

In addition, there were a number of smaller government-military computer projects involving mostly large corporate capital. Raytheon, for instance, under contract with the National Bureau of Standards (NSB), produced the RAYDAC (1947-51) on behalf of the Office of Naval Research (ONR); Radio Corporation of America (RCA) produced the Typhoon for the Navy in 1947 and, later, in 1955, the BIZMAC for the Army; General Electric produced the OARAC computer for the ONR in 1953 and throughout this period Burroughs was also developing computers under military contracts, delivering the ATLAS computer to the USAF in 1955 (Katz and Phillips, 1982).

Looking at the contemporary environment, it is clear that the Korean War greatly spurred the involvement of corporate capital in electronic computers. Indeed, IBM's decision to enter the field at all is said to have been prompted by the opportunities opening up with the conflict.[15] IBM proposed its Defense Calculator to government contractors in 1951, completing it in 1952 under the new name IBM 701. It was to be the first of a highly successful endeavour which was to turn the company into undisputed leader in this industry.

A final line of interaction between constituents involved supporting the emerging computer capabilities by backing research projects and new companies. This was particularly evident during the late 1940s and early 1950s as those involved in wartime projects not only strove to continue computer development, but also launched commercial ventures directed mainly to government and military markets. One example of this was the Institute for Advanced Study (IAS), organised by Von Neumann in 1946 with the participation of RCA Labs and funded by IAS and the Army and Navy Ordnance Departments. IAS is said to have been instrumental in the development of electronic computers (Soma, 1975). In terms of commercial ventures, government and military backing enabled the start of Engineering Research Associates (ERA) in 1946 and Eckert-Mauchly Computer Company (EMCC) in 1947. These were instrumental in developing the first commercial computers, the ERA 1101 and the UNIVAC. However, both lacked the financial resources to survive in the market, and were bought by Remington Rand (later, Sperry Rand) in 1952 and 1950 respectively [Pugh (1984), Katz and Phillips (1982)]. Computer Research Corporation and Electrodata Corporation experienced a similar fate, both founded in the early 1950s, again through government/military contracts. The first was bought by National Cash Register (NCR) in 1953 and the Electrodata Corporation by Burroughs in 1956, thus continuing the trend of large corporate capital dominating the field.[16]

By the second half of the 1950s, the computer industry was already heavily concentrated in the hands of large corporate capital whose influence within the technology's social constituency had greatly increased. Table 1 shows the share of the market for the firms which have dominated the computer industry between 1955 and 1971. As it shows, since 1956 IBM has dominated the industry, with an average share of 72.2% for the period 1956-71. In addition, the control by the four top companies has been consistently over 85% of the market for the same period (see also Sharpe, 1969). On the other hand, Figure 12 shows the variations in relative influence within the social constituency of computer technology through the changes in the relative share of the military-space market, in contrast with the commercial market. Clearly, until 1954, the military's dominance was overwhelming. This was the year when the first commercial computer, a UNIVAC, was delivered by Sperry Rand to GE (Schnee, 1978).

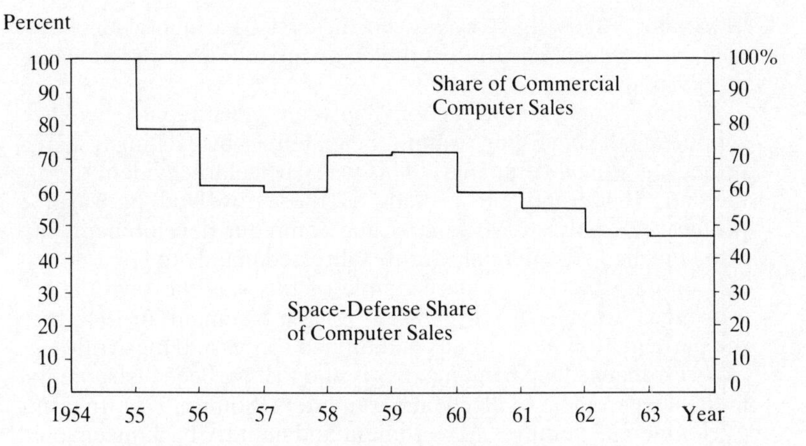

Figure 12. Percentage Distribution of Computer Sales Between the Commercial and Space-Defense Sectors (1954-63)
Source. Based on figures given in J. Schnee (1978). p.9

Table 1: Market Shares of Leading Computer Companies (1955-71). Percent

Year	Sperry Rand	Honeywell	Control Data	General Electric	RCA	Burroughs	NCR	IBM[1]
1955	38.5				5.1		0.3	56.1
1956	18.6				1.6	4.4	0.1	75.3
1957	16.3	.3			.8	3.9	0.06	78.64
1958	16.3	1.0		0.2	1.8	3.3	0.04	77.36
1959	17.8	1.2		0.9	1.4	4.2	0.12	74.38
1960	16.2	0.9	1.0	2.8	2.4	3.4	0.4	72.9
1961	15.5	2.0	2.2	3.4	3.0	2.6	0.7	70.6
1962	12.4	2.3	3.1	3.7	3.5	2.2	1.9	70.9
1963	11.2	1.8	4.0	3.5	3.5	2.6	2.7	70.7
1964	11.8	2.5	4.4	3.3	3.0	3.1	2.8	69.1
1965	12.1	3.8	5.4	3.3	2.9	3.6	2.9	66.0
1966	11.3	5.2	5.3	3.5	2.7	3.0	2.4	66.6
1967	10.6	4.7	4.7	3.0	3.2	2.9	2.5	68.4
1968	5.6	4.1	3.9	3.2	2.4	2.1	2.2	76.5
1970	3.2	4.8	7.3	3.1	2.1	3.4	2.3	73.8
1971	4.4	7.6	7.7		2.1	4.1	2.5	71.6

Source. G. Brock (1978), p.22
[1] Calculation from other figures. It assumes the share of any other company than those listed in the Table as negligible

By 1957, the role of the military was declining fast as commercial computer sales climbed up to 40%. However, this was the year of the Sputnik and the start of the space race which heightened the

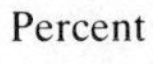

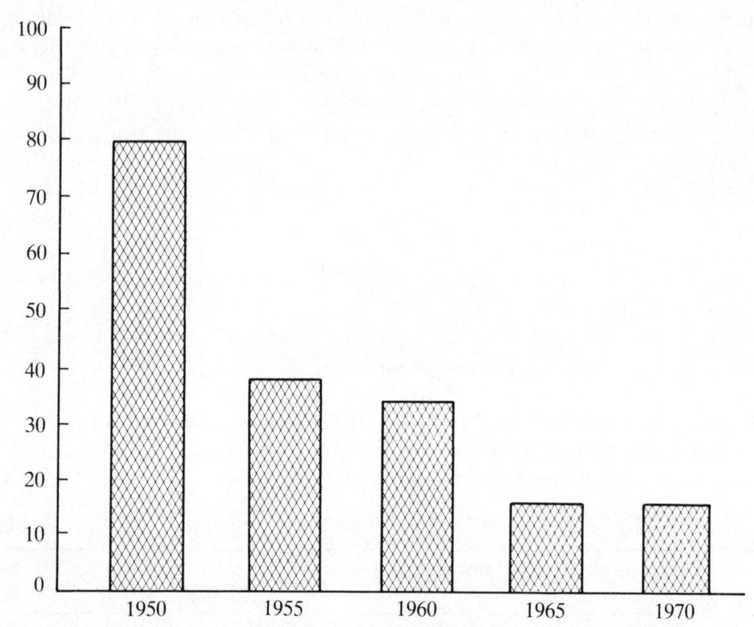

Figure 13. Percentages of Total Value of General-Purpose Digital Computers Installed in Government Agencies (1950-70)
Source. W. Sharpe (1969), p.194

pressures of war and international politics. As a result, in 1958 commercial computer sales fell by more than 10% while space-defense sales leapt by more than 40%.[17] In 1959, the space-defense share of total sales reached a peak of 72% before starting to decline as commercial computer sales once again gathered momentum. By 1963, the military-space share had finally dropped below the fifty percent line, thus heralding the end of its dominance in computer development. Figure 13 illustrates this with the case for general-purpose digital computers. By the mid 1960s, the value of digital computers installed by government agencies was less than 20% of the total. This decline was also reflected in research and development support. Whereas in the 1950s government-contract R&D accounted for about 60% of total expenditure at IBM, by the early 1960s such support had become only a small part of the company's R&D budget (Freeman, 1974). The government-military role was still important for the development of powerful computers incorporating advanced features such as the STRETCH. But as Freeman pointed out, 'the highly successful transistorized 1401 series (1960), the 360 series (1965) and the 370 series were private venture projects' (Freeman, 1974, p.133). In short, by the mid 1960s, large corporate capital seems to have become the dominant factor shaping the development of the United States computer technology, while the role of the government and the military, though still substantial, was gradually declining.

c) Semiconductors

The case of semiconductor technology and its industry is very similar to that of the computer. Although differing in details, the same nexus of social interests, playing much the same roles, is involved in its rise and development. The same lines of interaction between different social constituents as were outlined in the case of the computer, are also found here, although the invention of the transistor by Bell Labs meant that the role played by university research and development is much less pronounced.[18] Below I shall concentrate on the distinctive features and developments which have shaped the unfolding of the technology and industry as well as the interrelations among the social forces forming its social constituency.

The transistor's invention, unlike the electronic computer, was not related to military purposes, although the work leading to its development certainly benefited from the wartime radar research on semiconductor crystal detectors.[19] The Bell System's primary reason for developing the transistor was the company's need for a semiconductor amplifier to replace and improve upon the vacuum tube used in the electronic switching of telephone calls [Brooks (1976), Mowery (1983)]. Bell Labs more than any other industrial research organisation had the necessary resources to produce the transistor breakthrough (Hogan, 1977).[20] Scholars agree that this fact had crucial implications for the development of semiconductor technology and industry, for Bell, interested in the rapid advance of the technology and simultaneously under government investigation for monopolistic activities, decided to pursue a very liberal patenting policy towards the transistor [MacDonald *et al.* (1981), Freeman (1974), Silk (1960)].[21] This led to a rapid diffusion of the new technology as Bell published and organised symposia to pass on the techniques involved [Tilton (1971), Braun and MacDonald (1978), Nelson (1982)].[22]

At this stage, the military had played no significant role. In fact, it is said that the Bell team was rather apprehensive about any attempt by the military to keep the technology secret [De Grasse (1984), Freeman, (1982)]. Such fears were unfounded, however, as the military also seem to have been interested in the rapid development of the technology for their weapon systems. Misa writes that, 'During the 1950s the military was also involved in a number of publicity efforts aimed at disseminating the new technology' (Misa, 1985, p.267). From the beginning, therefore, the diffusion of semiconductor technology was unhindered by patenting as Bell's practice of liberal licensing subsequently became common practice for the industry as a whole.[23] It can be argued that much of the semiconductor industry's dynamism and its smaller degree of concentration, in comparison with computers, stem from this.

In practice, however, it was the involvement of government and military interests on one hand, and the interaction of semiconductors with electronic equipment industries (e.g. computers and communications) on the other, which produced the complex of social forces shaping the emergence and development of the US semiconductor industry. The role of the military was particularly important in the technology's early days in the 1950s, and in the crucial innovations that laid the foundations for the development of present-day microelectronics.

The electronic valve had proved to be highly unreliable in military equipment so that, in the wake of the Korean War, in 1951 'the three services assigned responsibility for military production development of the transistor to the Signal Corps' (Braun and MacDonald, 1978, p.80). The aim was to increase the availability of transistors, reduce their cost and improve performance (*ibid.*). This was achieved by a major injection of resources into R&D, and by helping to set up production facilities, as well as by procurement which clearly specified military needs, thus ensuring a stable market offering premium prices [Wilson *et al* (1980), Utterback and Murray (1977), Misa (1985)]. As a result, the technology advanced rapidly, very much in line with military needs.

By 1952, large established corporations (i.e. Western Electric, GE, Raytheon, RCA and Sylvania) were the first to benefit from the military funding allocated for setting up pilot semiconductor production lines [Schnee (1978), Levin (1982)]. According to Linvill and Hogan (1977), $11 million were invested in building alloy (germanium) transistor lines. Table 2 shows funds directly allocated to firms for semiconductor research and development and production refinements from 1955 to 1961. In particular, support for R&D had increased substantially since 1957-8. This undoubtedly reflected the space race and the shift towards an increasingly

Table 2: US Government Funds Allocated Directly to Firms for Semiconductor R & D and for Production Refinement Projects (1955-61)
Millions of dollars

Use of funds	1955	1956	1957	1958	1959	1960	1961
Research and development	3.2	4.1	3.8	4.0	6.3	6.8	11.0
Production refinement							
Transistors	2.7	14.0	0.0	1.9	1.0	0.0	1.7
Diodes and rectifiers	2.2	0.8	0.5	0.2	0.0	1.1	0.8
Total	8.1	18.9	4.3	6.1	7.3	7.9	13.5

Source: J. Tilton (1971), p.93

missile-based strategy (OECD, 1968b), which stimulated the miniaturisation programmes and led eventually to the invention of the integrated circuit. On the other hand, 1956 shows $14 million spent on transistor production refinements, reflecting the silicon transistor's impact and the need to build production lines for the new diffused-base process of production (Linvill and Hogan, 1977).[24] Direct Federal funding, however, does not account for all the backing received by the semiconductor industry. In terms of R&D alone, a 1960 Department of Defense survey found that in 1958 and 1959, direct and indirect funding amounted to $13.9 million and $16.2 million respectively; these figures were almost three times those for direct funding shown in Table 2 (Tilton, 1971). Industry certainly invested even more money in semiconductor R&D and production.[25] For instance, in 1958 and 1959, the government's share of the total expenditures in R&D was 25% and 23% respectively (*ibid.*). Nevertheless, military funding played a pivotal role in semiconductor development through the flexibility and timing of these investments. Funds were made available at critical stages, first, for the germanium transistor, then for the silicon transistor and later for the IC (Levin, 1982). Industrial interests most favoured for R&D appropriations throughout the fifties were the large corporations with well-established R&D facilities. By 1959, Western Electric and eight other large valve-producing electronic companies were receiving 78% of the government R&D funds, although their share of the total market was only 35%. New firms, which by then held 63% of the market, were receiving only 22% of government R&D funds.[26] For this reason, as Braun and MacDonald state, 'The semiconductor industry as a whole benefited from Government R&D expenditure and more from government production improvement funds, but the older firms benefited very much more than the newer' (Braun and MacDonald, 1978, p.81).

There is little doubt that through their R & D and production refinement programmes, the military social constituent greatly helped to shape semiconductor technology development in accordance with their individual interests. For instance, early funding to Bell Labs and others for pilot production was undertaken principally to provide the military with devices that could be used and tested in its communication equipment and weaponry (Levin, 1982). Likewise, the production of a silicon transistor was very much in the interests of the USAF, since devices capable of operating at higher levels of temperatures and radiation were needed for missile guidance systems. Also, according to Misa, transistors capable of amplifying high-frequency signals were urgently needed by the military to satisfy requirements in high-frequency radios, high-speed data transmission equipment, and high-speed computers. The development of the diffusion technique which enabled high-frequency transistors to be manufactured was clearly influenced by such a military need. In fact,

> The Signal Corps' support of diffusion research at Bell was striking . . . in the late 1950s, the Signal Corps' support of fundamental transistor development exceeded Bell's own in-house support in only a few cases, and all of these were devices produced by the diffusion process (Misa, 1985, p.281).

In the case of the silicon transistor, the military impetus was equally strong. Indeed, soon after the production of germanium transistors had begun in the early 1950s, the USAF launched a $5 million R & D programme on silicon devices. Texas Instruments (TI) were well aware of this military demand as they forged ahead with a research programme which was to produce the first silicon junction transistor in 1954.[27] The rewards were enormous: the silicon transistor was bought in bulk by the military and TI had the market for itself for about three years.

The invention of the integrated circuit by TI in 1958[28] again showed supportive military interests behind semiconductor technology. As Mowery describes

> As radar and military electronics increased in importance and complexity during the early 1950s the military services became more desirous of modular circuit designs that could reduce the number and complexities of interfaces within missile guidance and detection devices (Mowery, 1983, p.187).

Then, the launching of the Soviet Sputnik fiercely stimulated the search for smaller and more reliable devices, with the military now pouring money into miniaturisation research programmes. At first, the armed services tried different approaches involving mainly the large established corporations. Thus, between 1957 and 1963, the Army Signal Corps spent $26 million on the 'micromodule' programme designed to stack and encapsulate transistors. Most of the work was carried out by Radio Corporation of America [Kilby (1976), De Grasse (1984)]. From 1958, the Navy backed the thin film approach designed to print a circuit and passive components onto a ceramic wafer (*ibid.*). Finally, from 1959, the USAF supported the molecular electronics approach which purported to build a circuit in the solid without reproducing individual components. The $2 million contract was given to Westinghouse [OECD (1968b), Braun and MacDonald (1978)]. Eventually all these approaches proved unfruitful and were abandoned, with the IC concept being first developed by Texas Instruments aiming at military interests and at the micromodule programme [Wolff (1976), Kilby (1976)]. Soon after, the military became heavily involved in IC as TI immediately sought their support for further development. In 1959, the company received $1.15 million from the USAF and the next year $2.1 million for production refinement (Levin, 1982). Military interests were further expressed in the shape of the technology as the R&D contracts, in particular, called for the development of integrated circuits capable of performing several specific functions and fabricated out of silicon, rather than the germanium used in TI's first demonstration models (*ibid.*).

In the early sixties, the USAF IC support programme was expanded to include other companies, while at the same time other agencies, NASA in particular, also became heavily involved.[29] In all, from 1959-64, it is estimated that government allocated $32 million to IC research and development, with the USAF accounting for 70% of the total (*ibid.*). According to Wilson *et al.* (1980), government R&D spending in semiconductors peaked in the mid 1960s with the NASA space effort playing a leading role.[30] After that, 'funding did decline in the late 1960s . . . During the 1970s government support for R&D subsided significantly' (Wilson *et al.*, 1980, pp.154-5).

In the industry's early years, however, as indicated, R&D and production facilities support was only a part of the role played by the government-military social constituents. Equally important was their market role, with procurement ensuring a stable demand for semiconductor components. This demand is thought to have deeply influenced the shape of the industry in several ways. In Tilton's view,

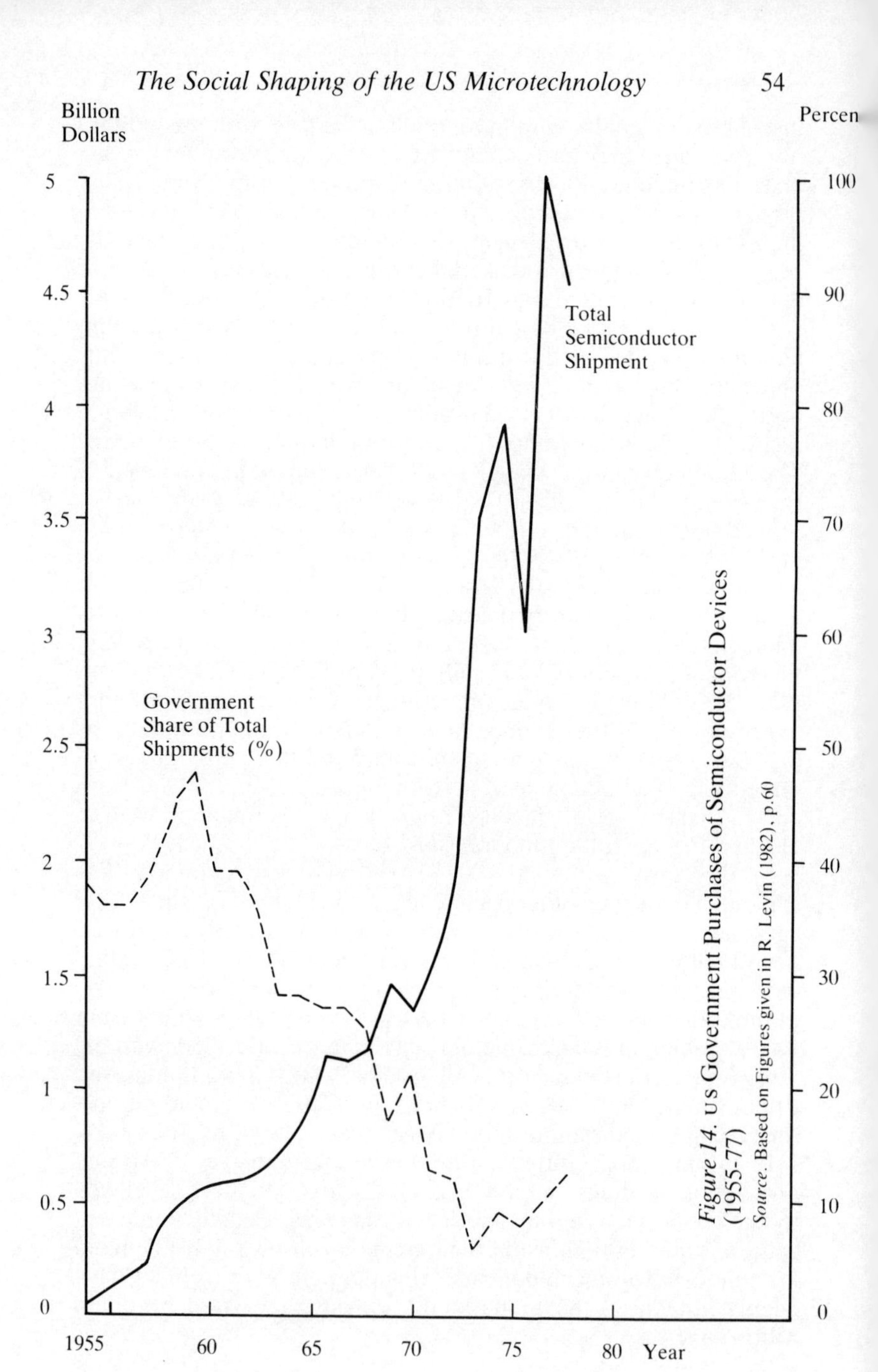

Figure 14. US Government Purchases of Semiconductor Devices (1955-77)

Source. Based on Figures given in R. Levin (1982), p.60

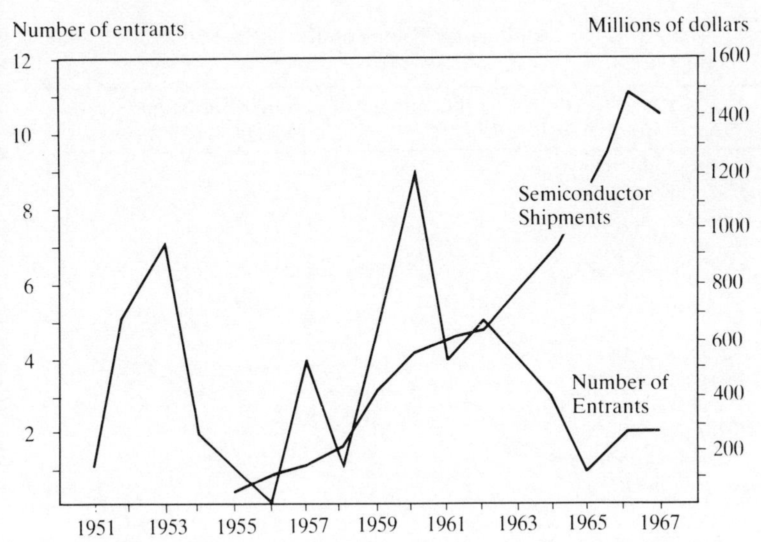

Figure 15. Entry into the Semiconductor Industry Compared with Growth in Semiconductor Production (1951-67)
Source. R. Wilson *et al* (1980), p.14

> Military demand . . . stimulated the formation of new companies and encouraged them to develop new semiconductors by promising the successful ones a large market at high prices and good profits. Further, the military market, by activating learning economies, often serves as a stepping stone to eventual penetration into the commercial market (Tilton, 1971, p.92).

Indeed, unlike the case of computers, many new firms were able to enter the semiconductor field by targeting their strategies towards the military who often provided the major or only market for their products, and at premium prices [Utterback and Murray (1977), Wilson *et al.* (1980)]. Figure 14 shows the importance of the military-space market for semiconductors between 1955 and 1977. Until the early 1960s, the average share was around 40% with a peak of 45% and 48% for 1959 and 1960 respectively,[31] reflecting the impact of the huge demand for semiconductors created by both the USAF Minuteman Missile programme and the NASA space programme. In comparison, Figure 15 illustrates changes in the number of entrants into the industry compared with growth in semiconductor production for the period 1951-67, while Tables 3

Table 3: Concentration of US Semiconductor Shipments: 1957, 1965 and 1972

Number of Companies	Percentage of Total US Shipments 1957	1965	1972
All semiconductors			
4 largest companies	51	50	50
8 largest companies	71	77	66
20 largest companies	97	90	81
50 largest companies	100	96	96
All companies	100	100	100
Integrated circuits			
4 largest companies		69	57
8 largest companies		91	73
20 largest companies		99	91
50 largest companies		100	100
All companies		100	100

Source: R. Wilson *et al* (1980), p.22

Table 4: Leading US Semiconductor Manufacturers (1955-80)

1955 Transistors	1960 Semi-conductors	1965 Semi-conductors	1975 Integrated Circuits	1980 Integrated Circuits
Hughes	Texas Instruments	Texas Instruments	Texas Instruments	Texas Instruments
Transitron	Transitron	Motorola	Fairchild	National Semiconductor
Philco	Philco	Fairchild	National Semiconductor	Motorola
Sylvania	General Electric	General Instrument	Intel	Intel
Texas Instruments	RCA	General Electric	Motorola	Fairchild (Schlumberger)
General Electric	Motorola	RCA	Rockwell	Signetics (Philips)
RCA	Clevite	Sprague	General Instrument	Mostek (United Technologies)
Westinghouse	Fairchild	Philco-Ford	RCA	Advanced Micro Devices
Motorola	Hughes	Transitron	Signetics (Philips)	RCA
Clevite	Sylvania	Raytheon	American Microsystems	Harris

Source: R. Levin (1982), p.30

and 4 give an idea of the changes in the industry's concentration and leadership from 1957-72 and 1955-80 respectively.

The pattern which emerges is one of concentration — although less than in the case of computers — accompanied by a relatively high degree of dynamism. New companies were able to benefit from the combined effect of rapid technological change, a liberal patenting environment and high military demand for new products under conditions that substantially reduced the risks of entering the industry.[32] Texas Instruments, for instance, came to lead the industry on the basis of the silicon transistor. Another company, Transitron, also succeeded on the basis of a new product whose work had initially begun at Bell Labs, the gold-bonded diode which was sold exclusively to the military (Tilton, 1971). In the late fifties, both Fairchild and Motorola entered the industry, achieving success by using silicon and relying on the oxide masking, diffusion, planar and epitaxial techniques that began in the early 1960s (Wilson *et al.*, 1980). Both received large contracts from the military.

Table 5: Average Price of Integrated Circuits and Proportion of Production Consumed by the Military (1962-72)

	Average price ($)	Percentage consumed by the Military
1962	50.00	100
1963	31.60	94
1964	18.50	85
1965	8.33	72
1966	5.05	53
1967	3.32	43
1968	2.33	37
1969	1.67	
1970	1.49	
1971	1.27	
1972	1.03	

Source: E. Braun and S. MacDonald (1978), p.113

Military-space encouragement of the IC was even more pronounced. As Table 5 reveals, initially, in 1962, when the product was expensive because its learning process was still in the early stages, the military-space share of the market was 100%. It was the USAF who opened the first major market, contracting TI to produce 300,000 integrated circuits for the Minuteman Missile programme. A family of twenty-two special circuits had to be designed and built [Layton (1972), OECD (1968b), *Electronics*

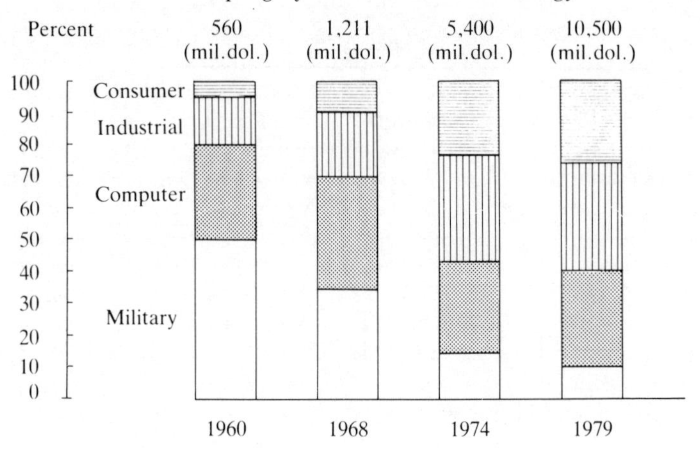

Figure 16. Distribution of US Semiconductor Sales by End-Use (1960-79)

Source. Based on figures given in R. Wilson *et al* (1980), p.19

(1980)].[33] NASA followed soon after demanding integrated circuits for its Apollo spacecraft guidance computer. The main contract went to Fairchild (Levin, 1982). In 1965, the space-military market for integrated circuits was still almost three-quarters of the total market and new contracts enabled companies such as Motorola, Signetics, General Microelectronics and Siliconix to enter into IC production (*ibid.*). By the late sixties, however, the military-space share of the IC market had dropped to 37%, reflecting both increasing demand from the commercial sector, particularly as a result of the third-generation of IC-based computers, and a decline in military-space expenditure, now that the Apollo programme was operational and a general reduction in military spending took place (Wilson *et al.* 1980).

Figure 16 shows that the same decline of the military sector occurred throughout the semiconductor field. From 50% of total semiconductor sales in 1960, it had fallen to 35% in 1968 and would continue to fall in the next decade. Thus, along with the R&D programmes, military-space orders during the fifties and first half of the sixties had decisively helped to shape the development of the technology and the industry. During this period, the government-military social constituents played a leading role and the commercial sector clearly benefited from it: not only from the numerous spillovers but, above all, from the innovation and acceleration of the learning process stemming from competition and

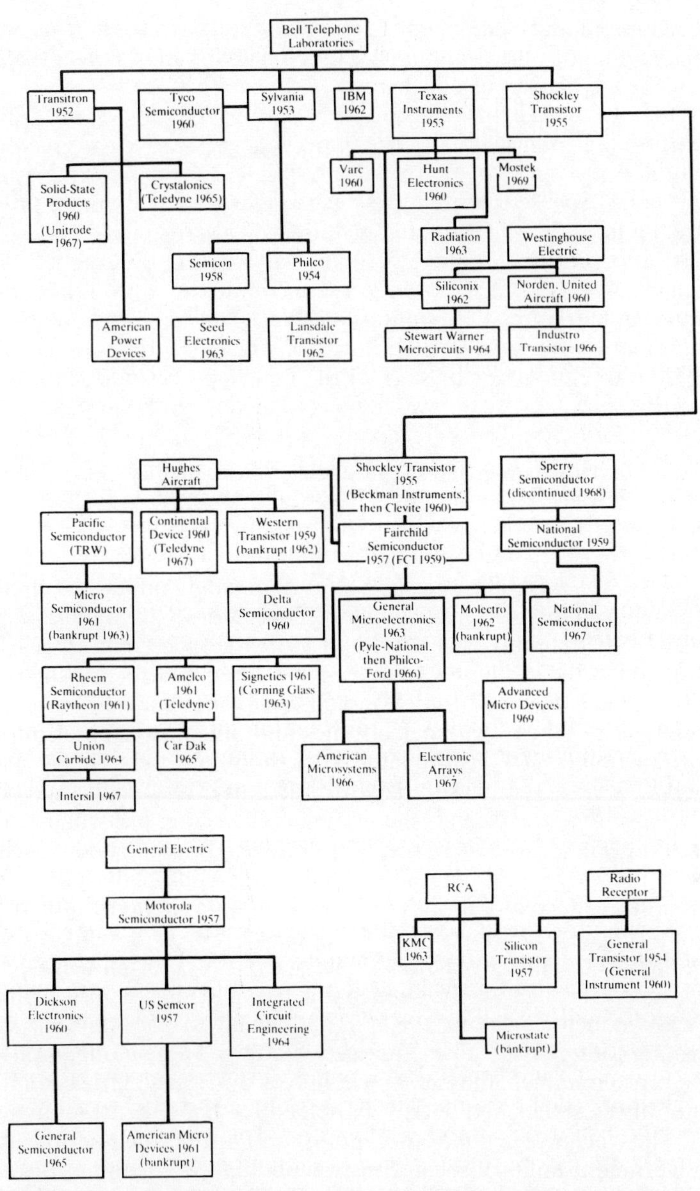

Figure 17. Spin-Offs from Bell Laboratories, Westinghouse Electric, Shockley Transistor, Hughes Aircraft, Sperry Semiconductor, GE, RCA and Radio Receptor *Source.* Č. Freeman (1974), pp.147-9

exacting military demands [34] which enabled the rapid diffusion of the technology into commercial use.[35] Table 5 shows this situation for the particular case of the IC.

Finally, unlike the case of computers, large corporate capital with well-established R&D programmes at the time of the invention of the transistor (e.g. Bell System, RCA, Raytheon, Westinghouse, Sylvania, Sperry Rand and GE) was no longer the dominant force by the late 1960s. To the factors already described was added the effect of antitrust policies,[36] like the Consent Decree of 1956 which allowed Bell to produce semiconductors only for its own internal market and the military market. This led to the industry's domination by the so-called new firms (see Table 4). Indeed, by 1959 the semiconductor sales of the new firms accounted for 63% of the total. As we know, however, in the same year the older established firms were receiving 78% of government funds for R&D and, overall, were spending 30% more than the new firms in semiconductor research and development. In the latter category, therefore, the role of older firms was still important, as was the part they played in forming new semiconductor firms.

It is a well-known fact in the industry's history that the origins of many of the new firms can be traced back to spinoffs from one or another of the large older corporations, particularly, the Bell System.[37] The family tree given in Figure 17 dramatically illustrates this situation. Thus, it seems clear that in its own way, large corporate capital was still a major social constituent in the early development of semiconductor technology. As will be seen, through mergers and acquisitions, its role has surfaced more clearly in the 1970s.

In conclusion, the development of industrial control, computers and semiconductors has revealed that, since the Second World War and until the late 1960s, the social constituency of microtechnology was generally the same as the social constituency of the United States research and development system. Consequently, the process of socio-historical developments affecting the unfolding of microtechnology followed a very similar pattern to what we have seen earlier in relation to the R & D system. Naturally, the specific form of each industry's development differed, for either technological or social reasons, but the fact of a government-military led social constituency stimulated by war and international politics, and largely free from international economic competition, underlies microtechnology's emergence and development. The exception is perhaps the telecommunications industry which had a long pre-war history of its own.

The Decline of Government-Military Influence

By the late sixties there was a marked change in microtechnology development in the US. The relative influence of government-military constituents waned as the emerging technologies and industries matured and corporate capital was able to exploit the vast opportunities offered by the commercial sector. In the wake of the Vietnam War and the beginning of the economic crisis, the decline in the military-space constituent's role began to be felt. As with the R&D system as a whole, this trend continued through the 1970s as the crisis deepened and the pressures of international economic competition heightened. The result was to give corporate capital a dominant place in microtechnology's social constituency. This was particularly illustrated by the case of semiconductors[38] where, as Figure 14 shows, the percentage of government purchases reached an unprecedented low of 6% in 1973, the year of the oil crisis, recovering to 12% in 1977. In turn, Figure 16 shows the computer, consumer and industrial markets accounting for roughly a third each of the distribution of US semiconductor sales.

In terms of R&D, the decline of the military role was equally dramatic. It was estimated that, by the late 1970s, government R&D funding comprised about 5% of the total R&D outlays for semiconductors (Wilson *et al.* 1980). For both market and R&D, therefore, by the late 1970s the military social constituent was no longer able to influence the development of microelectronics as it had done in the early days of the technology. Indeed, the most important breakthrough of the 1970s, the microprocessor, was developed entirely for commercial applications by the US company Intel (Sciberras, 1980). The spillover effect had swung dramatically from a military to a commercial origin. According to Mowery (1983), very little transfer of technology from the military to the civilian sphere has taken place since the late 1960s, while military applications of integrated circuits during the 1970s drew upon custom circuits similar to those designed for the civilian market.[39] The enormous importance of the civilian markets and the changing international competitive environment had apparently driven a wedge between the interests of military and corporate capital. At least this was the case as far as predominantly military interests were concerned. As two commentators have put it, 'The interests of the Military in high reliability and performance, whatever the cost, can no longer be those of the industry, if it hopes to compete in the commercial markets' (Braun and MacDonald, 1977, p.141).

By the mid-to-late 1970s, the social constituency of microelectronics was again a reflection of what we have seen in our analysis of the social constituency of the R & D system. That is, the commercial interests of corporate capital had become dominant amid a general slackening in the pressures of war and international politics, largely as a result of the Vietnam War and the economic crisis. This in turn had led to a loosening of interrelations between the dominant constituents associated with the pressures of war, to the point where their influence on the technological process had weakened considerably. This last fact was seen in the dramatic decline of military influence but was also confirmed by the sharp fall in (basic) scientific activities related to semiconductor technology. According to one analyst by the late 1970s such activity, particularly at universities, was insignificant, a trend seen by many as threatening the long-term development of the industry (Wilson *et al.*, 1980). All this was about to change however; military pressures began to revive while, at the same time the answer to the United States' international economic problems was seen to rest on science-based technologies — of which microtechnology is the most prominent. I shall now turn to this particular process which represents the current position of the social constituency of US microtechnology.

FOUR

THE CURRENT DEVELOPMENT OF MICROTECHNOLOGY I

The main features of the socio-historical background to the development of US microtechnology in the late 1970s and 1980s have been discussed earlier in relation to the US's research and development system. There we saw that by the late 1970s, economic and politico-military pressures were paving the way for a renewed convergence of interests between the constituents of the postwar government-military-capital-science social complex of power. This was the result of the long-term relative decline of the US economy in comparison with Japan and Europe, particularly West Germany, and the alleged decline in military power, following the detente policy and falling military expenditure in the 1970s. The path ahead, however, was not as straightforward as it had been immediately after the Second World War when the same complex of social interests was reconstituted, largely under government-military influence, with little economic competition from Europe and Japan. In fact, in the current phase we saw that the impetus of war and strong international economic competition may actually have conflicting results as both make claims on the resources of the R&D system and science-based technologies such as microtechnology. In the 1980s, therefore, the social constituency of microtechnology seems less cohesive than in the 1950s. Indeed, the entire social process underlying the development of microtechnology has become much more complex as military, economic, political and technical developments interact in a single sociotechnical process of national and international dimensions. The main trends of this process in the case of the social constituency of US microtechnology are discussed below.

It is possible to distinguish four sub-processes interacting significantly with each other.

a) Re-strengthening links between corporate capital, government, the military and science on the basis of war and global economic and technological competition. This has been seen most clearly in the proliferation of government-supported technological programmes and a variety of institutional arrangements seeking to strengthen links, for instance between government, industry and university, or between the military, industry, and university etc.

b) Closely related to this process is the military effort to regain influence within the social constituency of microtechnology through massive and strategically-placed use of financial resources, with the aim of reshaping a large part of the technological process towards military ends. A return to the military spillover mechanism is thus taking place.

c) Convergence and synergistic development of electronics. This has not only brought hitherto unrelated industries into collision but has also created the conditions for a huge process of vertical and horizontal integration within the electronics industry, as well as bringing large corporate capital from other economic sectors into the field.

d) Intensified electronics competition on a global scale. In a cause-and-effect relation with process (c) above, this has stimulated the capitalist tendency to create a market dominated by a few firms, at national and international levels. It has sparked national and international integration and/or coalition of capital at both intra- and inter-industry levels.

Let us see the practical manifestation of each of these interrelated processes, two of which we consider in this chapter and two in the next.

Re-strengthening Links Between Corporate Capital, Government, the Military and Science

This process was taking shape by the late 1970s as the overriding interests of corporate capital, government, the military and science were able, once again, to converge as the Vietnam crisis subsided[1] and the need was felt for a science-based technological response to the military and technoeconomic challenge facing the United States. As a result, industry, apart from its own research, has increased its support to university research[2] with funds rising from $84 million in 1973 to $370 million in 1983 (Nelkin and Nelson, 1987); in turn the university has itself encouraged links with industry, while government and the military have stimulated and participated in these linkages, and the military, in particular, have evolved their own technological programmes in an attempt to direct an important part of the technological drive of the renewed social constituency towards their own interests. All the social constituents emphasise the mutual and national benefits of re-strengthened linkages. Where tensions arise from their different institutional cultures, these are largely resolved with an eye to accommodating their main interests. In the early 1980s, these overriding interests, in the case of the university-industry linkage were described as follows:

> industry facing sharp international competition is looking to university research for some of the answers to its lagging productivity and technological innovation . . . money is what academia needs desperately right now (Stanfield, 1980, p.2021).[3]

In practice, this complementarity has been more important [4] and Table 6 gives an indication of the type of practical arrangements increasingly agreed. According to Nelkin and Nelson, in every case these arrangements 'involve an element of faith that they will be good for business, helpful and appropriate to universities and in the public interest' (Nelkin and Nelson, 1987, p.68). In fact, as indicated earlier, the government and the military have encouraged most of the important arrangements as a strong university-industry link is seen as the key to the acceleration of science-based technology and hence in everyone's interests.[5] Because of its strategic importance, microtechnology is perhaps the technology which most clearly shows the convergence of these interests. A large network of institutional arrangements has emerged which is looking to advance, not just mould, microtechnology's development.

The great push came initially from the government when, in January 1978, the National Science Board agreed to change the guidelines for research grants awarded by the National Science Foundation (NSF) in order to permit funding of co-operative research projects involving both university and industry (Dickson, 1984). Two years later, a bill was signed by President Carter authorising the expenditure of $285 million by the NSF and the Department of Commerce to establish industrial technology centres at universities.[6] Industry was expected to contribute at least 25% of each centre's funding and the latter were 'expected to become, ultimately, either self-sustaining or totally industry-supported' (*Physics Today*, December 1980, p.55). The military joined the push in 1983 by adopting a plan which would give a bonus on

Table 6: Types of University-Industry Links

Corporate contributions to university	Co-operative Research
Undirected corporate gifts to university fund.	Co-operative research projects: direct co-operation between university and industry scientists on project of mutual interest; usually basic non-proprietary research. No money changes hands; each sector pay salaries of own scientists. May involve temporary transfers of personnel for conduct of research.
Capital contributions: gifts to specific departments, centers, or laboratories for construction, renovation, equipment.	Co-operative research programs: industry support of portion of university research project (balance paid by university, private foundation, government); results of special interest to company, variable amount of actual interaction.
Industrial fellowships: contributions to specific departments, centers, laboratories as fellowships for graduate students.	Research consortia: single university, multiple companies; basic and applied research on generic problem of special interest to entire industry; industry receives special reports, briefings, and access to facilities, for example.
Procurement of Services	**Research Partnerships**
By university from industry: prototype development, fabrication, testing; on-the-job training and experience for students; thesis topics and advisers; specialized training.	Joint planning, implementation, evaluation of significant, long-term research program of mutual interest and benefit; specific, detailed, contractual arrangement governing relationship; both parties contribute substantively to research enterprise.
By industry from university: education and training of employees (degree programmes, specialized training, continuing education); contract research and testing; consulting services on specific, technical, management problems.	
Industrial associates: single university; usually multiple companies; industry pays fee to university to have access to total resources of university.	

Source: D. Prager and G. Ommen (1980), p.381

research and development contracts, to companies that worked with universities (*The Chronicle of Higher Education*, 4 May 1983). This idea of the military encouraging ties between industry and university had been proposed in 1980 by the government to the heads of the Army, Navy, Air Force and the DoD Advanced Research Projects Agency (DARPA)[7] (*Physics Today,* December 1980). Soon centres began to spread in the microtechnology field[8] as electronic corporate capital and universities converged.[9]

At Cornell University, for instance, a $5 million grant from NSF enabled the establishment of the National Research and Resource Facility for Submicron Structures. Of the eight members of the Facility's policy board, four came from major electronics companies, namely, Bell Labs, IBM, Hewell-Packard and Intel Corporation. In addition, through its affiliate programme, other major companies such as General Electric, Texas Instruments, Rockwell International and Xerox are also linked to the Facility (Dickson, 1984). Another major centre is the Center for Integrated Systems (CIS) at Stanford University, promoted independently by the university and bringing together corporate capital and the military, with a support of $12 million and $8 million respectively. Here major corporations were to give the university an unrestricted grant of $1,750,000, spread over three years, to help pay for a new building and research equipment,[10] while the military, through DARPA, were to help establish a facility at CIS for rapid fabrication of large-scale integrated circuits (Norman, 1982). MIT has followed Stanford with its twenty-million dollar Microsystems Industrial Group (MIC). Again, large corporations are injecting $750,000 over three years, but small companies can also join MIC for a $50,000 fee each. At Rensselaer Polytechnic Intitute (RPI), a thirty-million dollar Center for Integrated Electronics (CIE) has also been established. Here, big corporations such as IBM and GE have committed large sums of money: $2.75 million and $1.25 respectively. The Microelectronics and Computer Technology Corporation (MCC), is one initiative launched by corporations themselves, a non-profit joint venture promoted by Control Data Corp. By bringing together computer and semiconductor companies, they can pool their resources and share the cost of long-range research (Marbarch *et al.*, 1985). Among the twelve major corporations that joined MCC are Honeywell, Motorola, RCA and Control Data. The companies donate scientists and researchers to MCC, loaning them for up to four years. MCC's budget is about $75 million a year with a staff of two hundred and fifty. It is based in Austin, Texas, after a competitive bid involving fifty-seven cities in which Austin's private donors, the state and universities put together a generous package of incentives (*ibid.*). Recently, MCC's fortunes have attracted a lot of attention both because the inter-company cooperation is threading along the sensitive path of

US antitrust laws and because it is considered a direct response by the US industry to the Japanese practice of industrial cooperation [Peck (1986), Fischetti (1986), White (1985)].

There are other forms in which the social constituents of microtechnology are liaising to further the technology's development. In some places, it has been the state-government which has provided most of the money for establishing electronics centres, mostly at state universities. This has been the case in Arizona, Minnesota and North Carolina. In Arizona, for instance, the state-government has put twenty million dollars towards the Center for Excellence in Engineering (CEE) whose core activity will be solid state microelectronics and computer science. Industry's financial participation in this case amounted to nine million dollars. In other instances there has been a multilateral arrangement which pools financial resources from electronics companies, with the aim particularly of promoting basic research at universities. This is the case of the Semiconductor Research Cooperative (SRC) initially suggested by IBM but now sponsored by the Semiconductor Industry Association (SIA). The goal of the cooperative is to establish eight to ten 'centers of excellence' at universities known for their work in key areas of microelectronics and computer science (Ploch, 1983). By 1986, total research funds distributed in this way was expected to reach $40 million a year. The first grants were made in November 1982, when $3 million were shared between the University of California at Berkeley, Carnegie-Mellon University (CMU) and eight other universities to support research on integrated circuits (Dickson, 1984). In 1986, SRC was supporting the research of about two hundred faculty members and four hundred graduate students with contracts totalling more than $15 million (Wallich, 1986).

Another form of university-industry linking is that pursued by CMU in relation to its Robotics Institute. Here the main founders are Westinghouse, DEC, and the Navy. Westinghouse provides $1.2 million and DEC $1 million a year for research projects. Unlike the cases above, the corporations will hold the patents arising from the research. In addition, the Institute has an affiliate programme with seventeen companies, each paying from $10,000 to $50,000 annually for periodic reports of work in progress and attendance to briefing sessions (Norman, 1982). More recently, a Software Engineering Institute (SEI) has been established at CMU which is being entirely financed by the military at a cost of $103 million. SEI will employ two hundred and fifty people by 1988, with the specific mission to 'accelerate the transition of emerging or advanced computer software technology into use in the development and maintenance of

DoD weapons systems' (*Datamation*, 1 October 1984). It is stressed, however, that SEI's activities will 'exclude development of mission software for defense systems' (*Electronics Week*, 4 February 1985). In turn the California Institute of Technology (Caltech) is running a successful affiliates programme with its Silicon Structures Project (SSP). Corporations pay Caltech $100,000 to send a staff scientist to the university to work for a year on computer-aided design of integrated circuits. Twelve firms were participating in this project in 1982. Finally, the NSF has also stimulated cooperative research projects carried out jointly by academic and industrial researchers. Here, corporations are required to supplement NSF's funds. Among the projects in progress, Bell Labs and Lehigh University are working on thermal convection in cavities; several computer firms are working with Caltech on design of silicon structures; Raytheon is working on signal processing with the University of Rhode Island and Westinghouse and the University of Florida are working on robot manipulation [Prager and Ommen (1980), Dickson (1984)].

In this way, the social constituents of microtechnology are fusing their interests behind the development of the technology, ultimately shaping it. There is a trend towards integration in scientific activities, particularly in generic research. This brings together not only the different social constituents of microtechnology but also competing corporations within the electronics industry constituent, different universities within the science constituent, and different departments within particular universities.[11]

The Renewed Influence of the Military on the Current Development of Microtechnology

Closely interrelated to the above process is another process in which the military is seeking to direct the development of key areas of microtechnology towards their own needs. Aware of their reduced influence within the social constituency of microtechnology, the military realise that unless they devise their own technological programmes they will have to rely on civilian spillover to meet their technical needs. Thus, by the late 1970s, efforts to redirect a significant part of the technological process began to take shape eventually crystallising in such huge programmes as the Very High Speed Integrated Circuit (VHSIC) and Sematech, the Strategic Computing Plan and the ICAM and MANTECH programmes for automation. Lately, the Strategic Defence Initiative project threatens to give the military a hold on the development of microtechnology which will continue well into the next century.

a) The VHSIC and Sematech Programmes

For the military, the basic tenets and rationale behind the VHSIC program[12] have been described by W. Perry, Undersecretary of Defense for Research and Engineering

> We have little ability to influence those companies whose sales are predominantly commercial. This is a serious limitation in the case of the semiconductor industry, whose products play a crucial role in nearly all of our advanced weapon systems. Therefore, we have initiated a new technology program intended to direct the next generation of large-scale integrated circuits to those characteristics most significant to Defense applications . . . This initiative will require expenditures of $31 million in FY 1980, and involve a total cost of about $200 million over six years . . . we expect this investment to stimulate at least an equal amount in industry . . . Our goal is to get the full benefit of our investment plus the added benefit of influencing the direction of a substantial amount of company R&D (W. Perry, quoted by Mowery, 1983, p.191).

The technological goal of the VHSIC program is to provide microcircuits in which high-speed, complex real-time signal processing is essential.[13] Thus, reducing the minimum feature sizes and increasing the functional complexity on a chip are only part of the programme. In addition, low power consumption is desirable and some specific military requirements, like radiation resistance, must be met [Connolly (1978), Sumney (1980), Hazewindus (1982), *Aviation Week & Space Technology* (1981)]. The programme initially divided into three phases, and the three Armed Services are cooperating in its development. Large corporations are the main or prime contractors and they are joined by other companies and universities and research institutes as team members. Table 7 shows how the diverse social constituents have teamed up behind different technical projects for Phase I of the programme. Table 8 shows the corporate and science constituents who will be involved in Phase III as well as the impressive array of technical projects which are involved in the VHSIC programme. Phase I was expected to end by May 1984. However, in 1983 it was already clear that not only was this phase going to go beyond this date but, also, that the whole programme was changing, with Phase II possibly becoming a Phase IV, thus extending the programme towards the late 1980s (Beresford, 1983). In the meantime, the costs related to Phase I were increasing, not only because the twenty-eight chips involved

Table 7: Capital-Military-Science Constituents Behind Phase I of the VHSIC Programme

Prime contractor & team members	Defense Department's VHSIC Program Phase 1: (Proposal) Technology	Architectural Approach	Phase Zero Monitor
General Electric Intersil Martin Marietta* Analog Devices Tektronix Standford Univ	C-MOS and C-MOS/SOS	Standard chip set of seven logic devices, three memory chips	Army
Honeywell	Bipolar (integrated Schottky logic, current mode logic, and current sourcing logic)	Standard families of chip sets and CAD	USAF
Hughes Aircraft Signetics Burroughs Res. Triangle Instit. Cornell Univ. Stanford Univ. UCLA	C-MOS/SOS and bipolar (integrated) Schottky logic	Not available	Army
IBM Northrop†	N-MOS	Standard chip set for data processing and macro cell standard custom via CAD	Navy
Raytheon Fairchild Semi-conductor Varian/Extrion	Bipolar (isolated integrated injected logic)	Standard chip set of four and custom 'glue chips' using gate arrays	USAF
Rockwell Sanders Assoc. Perkin-Elmer	C-MOS and C-MOS/SOS	Standard chip set: four logic plus two memory	Army
Texas Instruments	Bipolar (Schottky transistor logic) N-MOS memory	Standard chip set of five and gate array 'glue chips'	USAF
TRW Motorola Sperry Univac GCA-Mann	Bipolar (three-diffusion) and C-MOS	15-20 chips total in two technologies	Navy

Table 7: Capital-Military-Science Constituents Behind Phase I of the VHSIC Programme *cont.*

Prime contractor & team members	Defense Department's VHSIC Program — Phase 1: (Proposal) — Technology	Architectural Approach	Phase Zero Monitor
Westinghouse National Semi-conductor Control Data Corp. Harris Electronics* Mellon Institute Boeing Aero-space†	C-MOS and C-MOS/SOS	Standard chip set of six and set of gate array 'glue chips'	Navy

* Joined team after Phase Zero award
† Proposed additions for Phase 1 effort
Source: Aviation Week & Space Technology (1981), p.50

Table 8: Social Constituents Behind Phase III of the VHSIC Program

Company	VHSIC Programs — Phase 3 Scope of Work	Duration (months)	Funding (thousand)
Amer. Science & Engineering	Development of a concentrating collimating illumination system for X-ray lithography	11	$600
AVCO	High intensity pulsed plasma x-ray source for microlithography	15	249
AVCO	Coronaphoresis for gas purification	9	220
Boeing	Storage/logic arrays	12	455
EBM Corp	Ultra-high-speed submicron direct write electron beam exposure system	20	1,243
GE	Development of MESFET silicon on sapphire for IC gates	12	338
GE	Improved performance package for VHSIC	18	235
Hewlett-Packard	Advanced resist materials and processes	30	491
Honeywell	Electronic packaging for VHSIC	25	470
Hughes	Low temperature photochemical processing for VHSIC applications	24	574
Hughes	Laser annealing	24	700
Hughes	Electron beam processing	24	392
Hughes	Refractory metal for interconnection	24	422

Hughes	Failure management design system	24	703
Hughes	Static induction transistor logic technology, low-temperature silicon epitaxy, and improved resists for electron-beam lithography	24	1,124
Hughes	Acoustic microscopy for inspection of VHSICs	28	602
Hughes	E-beam lithography components for directly writing very high-speed integrated circuits	24	1,541
Hughes	Electron beam circuit tester	18	451
Lockheed	A study of VHSIC applications for naval patrol aircraft	14	551
Perkin-Elmer	Extend microlitho technology through plasma etching research	24	378
Perkin-Elmer (Harris, Mead)	Proposal to extend microlithography technology	15	1,520
Raytheon	High density performance hybrid circuit	18	180
Rockwell	Low resistivity gates for C-MOS integrated circuits	17	95
Sanders	Memory processor study	15	278
TI	Data flow architecture	10	436
TRW (GCA)	Electron beam system software	15	1,369
TRW	Transportability of CAD data	10	479
TRW	Software architecture study	12	754
Varian	Development of direct write E-beam lithography system	48	1,215
Westinghouse	Electronically alterable ROM for VHSIC	24	581
Westinghouse	Low defect density silicon substrates for NMOS	24	599
Westinghouse	Mobility and drift velocity measurements in inversion layers	14	83
Westinghouse	Improved SOS for VHSIC	24	248
	Universities & Research Institutes		
Univ. of Arizona	Signal processing algorithms on chips	12	131
Carnegie Mellon	A hierarchical design approach for VHSIC	24	389
Cornell Univ.	Studies on laser annealing of SOS and poly si/insulator	36	282
Cornell Univ.	Simple submicron device models for circuit simulation	36	256
Cornell Univ.	Exact simulation of submicron scale devices	36	256

Table 8: Social Constituents Behind Phase III of the VHSIC Program *cont.*

VHSIC Programs — Phase 3			
Company	Scope of Work	Duration (months)	Funding (thousand)
Cornell Univ.	Analytical methods for determination of substrate defects introduced during processing	36	204
USC	Design automation	24	281
USC	Design for testability and reliability	24	332
Univ. of Illinois	Reliable, high performance VHSIC system	48	2,696
Research Triangle Institute (RTI)	Identification and assessment of on-chip self-test and repair concepts	12	170
RTI	Design, development and application of a signal processor architecture performance evaluation tool	48	519
SRI	Fault tolerant architecture for VHSIC	12	251
SRI	Intense plasma x-ray source for submicron lithography	33	493
SRI	Assignment algorithms for the control of VHSIC semiconductor chips	12	100
Stanford	Testing VHSIC devices	36	530

Source: Aviation Week & Space Technology (1981), p.54

were delayed, not arriving until the end of 1985 (nearly half had appeared by the end of 1984) but, also, because two additional programmes have been devised to ensure the rapid insertion of the new chips and an improved yield during their production. Because of the insertion programme, by the end of 1984 the Office of the Under Secretary of Defense for Research and Engineering, which is managing the VHSIC programme, had already seeded thirty-eight insertion programmes at a cost of about $150 million, and an additional $50 million in seed money was being planned. On the other hand, the VHSIC-sponsored yield-enhancement programme involving all the Phase I contractors had meant another $90 million from the office. In turn, the three services had put up an additional $300 million to support insertion contracts (Iversen, 1984). As to the entire VHSIC programme, its cost was put at $680 million, excluding the insertion and yield-enhancement programmes related to Phase I (*ibid.*). Thus, as frequently happens with military projects, the actual cost of the VHSIC programme has in the end meant far more resources being devoted to military purposes than originally proposed. Thus the following statement should be read with some apprehension. It suggests that DARPA, 'in an effort that is separate but complementary to the VHSIC programme, is looking into the future when it may be possible to fabricate more than a million gates on a single chip' (*Aviation Week & Space Technology* 1981, p.54). For all this, it seems plain that the military are aiming for a long-term shaping of semiconductor technology in line with their own interests. The VHSIC programme has already given them a strong foothold for the renewal of this influence. But more is likely to follow, for in this process, they are being favoured by the US semiconductory industry's decline in the world market which has forced the industry's leaders into an anxious search for ways of reversing such a trend.

The current difficulties of the US semiconductor industry will be discussed later. Here my main concern is simply to emphasise how such difficulties are leading, almost inevitably, to a strong participation of military interests in the development of the industry in the absence of other important civilian-oriented government programmes. A brief look at the way in which the recent Semiconductor Manufacturing Technology (Sematech) project is taking shape will confirm this point. As proposed by the US semiconductor industry, Sematech is a $1 billion government-industry consortium, intended to develop advanced semiconductor manufacturing techniques for US chipmakers in an attempt to restore their international competitiveness and leadership (*Electronics,* 5 and 19 March 1987). In an unprecedented degree of collaboration which clearly tests the US antitrust laws, the consortium seeks to bring together the expertise and funds of not only US chip producers but also their US suppliers and major customers such as US computer makers. This is a concept that takes the research and development collaborative ventures, seen in the previous section, a great deal further, almost onto the market place. The plan includes the generation of an advanced production facility which will produce enough chips to make sure of the efficiency of the manufacturing processes. The original plan envisaged a high-volume commercial production facility, but this has been scaled down by the US Semiconductor Industry Association (SIA) in a decision which favoured the Sematech concept supported by IBM and Intel. Sematech was given the Congressional go ahead late last year and is expected to produce devices by the end of 1988. Congress's approval for the programme amounts to an initial $100 million for the fiscal 1988's budget. A similar amount must be contributed by the partners in the consortium. At this point and not unexpectedly, the military has already stepped in as the source of government funds, which means that they will have a guaranteed say in the future of Sematech and hence, in the future of US semiconductor technology. The way the process has been handled reveals the determination with which the military is prepared to move into developments which it sees as fundamental for its own advancement.

First, the military has supported the case of the US semiconductor industry giving due emphasis to implications for defense and national security, something that the industry itself had been forcefully arguing. This position was made official in a report by the Defense Science Board task force which, not coincidentally, was released only a few weeks in advance of the industry's own Sematech proposals and recommended the establishment of a very similar 'semiconductor manufacturing technology institute' to be run by a consortium representing US semiconductor producers (*Financial Times*, 27 January 1987). The report also recommended that the Department of Defense should support the capitalisation of the institute with $200 million per year for five years (*ibid.*). In short, the military has reacted to answer the semiconductor industry's anxiety for government support, thus ensuring their participation in the industry's development. As shown below, however, this kind of development has brought echoes of the old controversy over the impact of the military on the United States R&D system. Questions have been raised about the impact of this renewed military influence on semiconductors when commercial markets and strong international competition, particularly from Japan, have achieved a major importance. [De Grasse (1984), Mowery (1983), Molina (forthcoming)].

b) The Strategic Computing Plan (SCP)

In the field of computers and AI, the military, through DARPA, launched their Strategic Computing Plan (SCP) in 1983. It was hailed by some as the US response to the Japanese Fifth Generation Computer Project and a return to the space race ethos. In the view of Botkin and Dimancescu,

> Twenty-five years go, DARPA led the charge to restore our technical leadership in the space race against the Soviet Union. The pieces were there but they had to be pulled together and focused on the mission. Now we are faced with another race — this time with the Japanese — for leadership in information processing. It took Japan's Fifth Generation Computer Project to awaken our competitive spirit now.

> Again DARPA has seized the initiative. Its strategic computing program with five-year funding of $600 million was recently submitted to the Congress. It lays out a multiyear research and development program to assure US superiority in information processing . . . The basis for this focused effort is national security, but the potential for strengthening our industrial base is also enormous (Botkin and Dimancescu, 1984, pp.224-5).

What the authors seem to overlook however, is not only that historical conditions have changed considerably since the space race, but that owing to its focus on commercial markets, Japan's interests are different from those of DARPA in the US. In the space race with the Soviet Union, their technological aims were similar, unlike the current race, where Japan is striving for social and industrial goals[14] and the US for military ones, expecting only indirectly to defeat the Japanese industrial challenge.

Looking at the Strategic Computing Plan's research goals confirms this point. Their three-pronged plan includes intensive efforts in microelectronics, computer architecture and AI software, all of which will be closely tied to specific military applications [Tucker (1985), Sun (1983), Ornstein *et al.* (1984)]. Among the highlights of SCP, it is envisaged that gallium arsenide chips will gradually replace silicon chips as a means of achieving faster computing, lower power consumption and greater resistance to nuclear radiation. In turn, work in computer architecture will focus on ways to run several computers in parallel, thus obtaining at least a thousand-fold increase in computing power. Finally, software research will put the emphasis on AI since it is anticipated that expert systems will play a vital role in the evaluation of complex problems and high-level planning while machines will be able to see, and to understand speech. Whether and when these goals will be achieved is still uncertain, but that such a programme exists is in itself a triumph for a military constituent seeking to shape microtechnology towards its own purposes.

SCP is offering a product to each of the three armed services. For the Army, a class of autonomous vehicles that could navigate eighty miles from one destination to another independently performing reconnaissance and other military missions. For the USAF, an automated pilot's associate which would be a crucial aid to aircraft operators facing increasingly complex aircrafts and demanding combat-flight conditions. For the Navy, an expert battle management system to help commanders of carrier battle groups plan and conduct major sea battles [Sun (1983), Tucker (1985), Ornstein *et al.* (1984)].

Clearly the military's interest in computers and AI is differently motivated from that behind the Japanese Fifth Generation Project. As DARPA's computer director, Robert Kahn, has put it, 'This is a very sexy area to the military, because you can imagine all kinds of neat, interesting things you could send off on their own little missions around the world or even in local combat' (quoted by Marbarch *et al.*, 1985, p.65). Thus, there is good reason to question how such military attitudes and consequent demands on the field of computers and AI can be construed as a direct response to the Japanese challenge. A connection between SCP and the Japanese challenge can only be found in the fact that large corporate capital may be able to manipulate the military demands for its own commercial purposes.

c) The ICAM and ManTech Programmes

In the field of industrial control and automation in general, the military has also taken the initiative, moulding the technology's development and involving corporate capital and universities in the effort. Following the tradition of the 1950s, the USAF started the first project, the Integrated Computer-Aided Manufacturing (ICAM), in 1977. The ICAM programme aims at developing computer-integrated manufacturing systems to be used within industry, in particular that related to the military sector (Hudson, 1982). Through ICAM, better productivity, lowered costs, greater design flexibility and greater management control over production were seen as attainable goals for the manufacturing industry. The first two aims, however, have already been seriously questioned [Noble (1984), Schlesinger (1984)], while for industry the last goal seems to offer the highest returns (De Grasse, 1984). Indeed, according to Noble, 'The Air Force Systems Command (AFSC) estimated a 54 percent reduction in "people" (blue-collar workers) as one of the chief Air Force CAM center benefits' (Noble, 1984, p.331).

Like VHSIC and SCP, ICAM also brings the convergence of corporate capital-government-military-science interests. Thus, between 1977 and 1982, the programme awarded more than sixty-five contracts involving US companies and universities and, between 1979 and 1984 expenditure was expected to reach $100 million. Currently, the AFSC has twenty ICAM projects under way, including a 'Factory of the Future' where a team of ten contractors, led by Vought Corporation and including North Carolina's Research Triangle Institute, is trying to develop an entire plant that uses only cutting-edge production technologies (Schlesinger, 1984). In addition, the AFSC is actively promoting a wide dissemination and transfer of this model factory between 1985 and 1990.

Most importantly, the ICAM programme has led to similar efforts from the other services as well as to a spin-off, the $67 million TECMOD (Technology Modernization) programme devoted to refining specific pieces of computer-integrated systems. The Navy's programme is the Shipbuilding Technology Program (STP) and the Army's is the Army Tank Command's Flexible Manufacturing System Project. In addition, there is the Tri-Service Electronic Computer-Aided Manufacturing (ECAM) project managed by the Army (Noble, 1984). All these projects have come to be seen as part of a military-led effort to refine manufacturing technologies called ManTech. Between 1977 and 1981, it is estimated that $640 million were invested by the military in ManTech and $141 million were used in the year 1983 alone [Tirman (1984), Schlesinger (1984)]. Currently, the Army is managing about six hundred ManTech projects worth a total of $300 million; and a 1982 Department of Defense projection for the following five years suggested that $1.5 billion would be spent on the automation effort. ManTech, therefore, has developed fully the ICAM concept of a military-led social constituency pressing for, advancing and shaping the development of the microtechnology of industrial automation. Not only is the scale of the programme much larger, but the social constituency itself has grown in strength, becoming more firmly institutionalised.

With ManTech, the military has become a major driving force behind automation, while organisations such as the industry's Manufacturing Technology Advisory Group and the academic community's College CAD/CAM Consortium have strengthened the links between military, the industry, and university (Schlesinger, 1984). Despite the fact that most investment focuses on weapon production, this social constituency presents itself as not only improving shopfloor efficiency for the defense industry but, in fact, as leading the US battle for international industrial leadership. As one commentator has said,

> DoD increasingly promotes its MT [ManTech] efforts as a boost to the country's global market-share and reputation for industrial leadership. Military MT is to drive *all* US industry, not just predominantly defense sectors, towards a more competitive posture (Schlesinger, 1984, p.189).

It is too early to say how far this is going to be so, but as in the case of VHSIC and SCP, indications are that the Japanese challenge will be equally strong in the field of industrial automation.

The three programmes I have described have undoubtedly heightened the influence of the military social constituent of microtechnology. The VHSIC, SCP and ManTech projects embrace almost every aspect of microtechnology (without taking into account the military involvement in the other major programmes such as the space Shuttle programme, with obvious implications for satellite communications), and their time-span promises the military a significant presence well into the 1990s. Furthermore, all these programmes are presented as leading the renewal of US economic and competitive strength, in case the commercial pressures weaken the relative influence of the military. This happened in the early 1970s when, following the Apollo Program, the Vietnam defeat, and the economic crisis, the military was left with no rallying cry to safeguard its interests. Today, VHSIC, SCP and ManTech are both military and economic programmes and their scale almost guarantees that military spinoffs will once again become an important aspect in the development of US microtechnology. It is in this context that we have to evaluate the likely impact of the latest military initiative, the Strategic Defense Initiative (SDI), first announced in 1983 and popularly known as 'Star Wars', since it is quite clear from its technical objectives that it will ensure the government-military constituents the role of driving force in the development of US microtechnology well into the twenty-first century.

d) The Strategic Defense Initiative (SDI)

In the descripton given by G. Yonas, chief scientist for the Strategic Defense Initiative (Yonas, 1985), the SDI programme is envisaged in four phases of which the first phase, from now until the early 1990s, will be devoted to research.[15] No system has as yet been decided, but four major elements have nevertheless been identified as crucial in any SDI system. These are a) surveillance, acquisition, trading and kill assessment; b) interception and destruction; c) battle management; d) survivability, lethality and key technologies. All these elements rely heavily on microtechnology and are bound to impose exacting demands upon its development, incorporating and taking even further the advances of VHSIC and SCP. Surveillance, for instance, will depend on advances in sensor technology, computers and the computer software to process the data generated. Interception will demand low-mass guidance and control systems for projectiles, capable of surviving initial acceleration for example.

But it is perhaps in the third element, battle management, that demands will be most stringent, for what is envisaged here is the integration of multiple layers of defensive weapons into a highly reliable and fault-tolerant system. In Yonas' words, this

> will require high-performance computers, sophisticated software and adaptable communications networks far beyond existing capabilities. The system will need to track tens of thousands of objects, from launch to destruction, and allocate defense weapons most efficiently. This defense system will require a network of space-based computers capable of performing millions of operations per second, surviving virtually maintenance-free for years in space and adapting to flaws and failures within the network. Equally exacting will be the requirements for software. Programs must be written and tested exhaustively to make sure that they are free from errors. In fact, the job of producing the necessary software may have to be done by advanced computers which would mean developing computer-controlled programming and debugging. (Yonas, 1985, p.31.)[16]

Programmes such as VHSIC and SCP seem mere preparations for SDI and the military shaping of microtechnology well into the twenty-first century. Thus, although a great deal of sceptism has been expressed from many quarters about the technical feasibility and strategic wisdom of SDI goals, particularly regarding its software,[17] research funding for 1985 was about $1.3 billion, with gradual increases manifested in a $5.9 billion request in SDI funds for fiscal 1988. For the first five years of the exploratory research programme the budget requested is $26 billion [von Hippel (1985), Brown (1985)], and in the medium term it may even grow to consume up to a fifth of US defense R&D funding (MacKenzie, 1985). As for the total cost of the SDI system, it has been estimated at around $800 billion, that is, as it has been compared, it is like 'adding another Navy to the budget' (Adam and Fischetti, 1985). Figures could go even higher if, as suggested by a recent study by the Congressional Research Service, 'Soviet countermeasures to a space-based missile defence system could boost the cost of deploying "Stars Wars" to as much as $1,000 bn' (*Financial Times*, 4 August 1987, p.4).

Tables 9a and 9b give an outline of the policy arguments for and against pursuing SDI. In the meantime, the military-

Table 9a: Points and Counterpoints in the SDI Debate: Arguments in Favour

Policy arguments in favour of pursuing the Strategic Defense Initiative		
Point	Counterpoint	Rebuttal
The SDI will make ballistic nuclear weapons 'impotent and obsolete'.	As a defense, the SDI will not be totally effective, hence, nothing can remove the the threat of nuclear war.	Without the SDI, the United States has no defense whatsoever. Since the SDI's layered defenses would prevent an adversary from knowing which nuclear weapons would get through, it would decrease their military effectiveness and preserve some chance of launching a retaliatory strike.
The SDI will lead to voluntary reductions in nuclear arms by both the United States and the USSR because it will rob ICBMs of their omnipotence.	Testing or deployment of the SDI would destroy the Anti-Ballistic Missile Treaty, and a major means of overcoming the SDI would be by multiplying missiles and decoys and building new types of missiles — all of which would mean breaking of SALT I and SALT II constraints, and little hope for new arms control.	The doctrine of mutually assured destruction is a shaky balance, in spite of the treaties. The number of nuclear weapons has mushroomed since World War II.
The SDI does not need to be perfect to be effective. Even an imperfect defense would help defend against attacks from accidental launches, terrorists, and 'outlaw' nations.	Even a defense that is 99 per cent effective is unacceptable because some people would be killed. Cheaper ways can be found to bolster deterrence. Nuclear missiles can be fitted with command-destruct devices to prevent accidental firing. Also,	In war or terrorism, violence and death are inevitable. A system that can reduce casualties by a factor of 10 is worthwhile.

	terrorists are likely to deliver a bomb not with an ICBM but with a suitcase, against which the SDI would be no defense.	
The SDI will strengthen deterrence by denying the USSR the confidence that it can achieve its military goals, because the SDI increases the survivability of US retaliatory forces.	On the contrary, the SDI might destabilize the nuclear balance by giving the United States the confidence to launch a first strike. An imperfect defense is much more effective against a ragged retaliatory blow than against an organized first strike.	A defensive system is designed to preserve peace, not promote aggression. Moreover, as history shows, in the 1950s and 1960s the United States had a clear first-strike capability, and it did not use 'nuclear blackmail'.
The SDI will protect US allies in Europe against a nuclear attack.	The Soviet Union would be able to attack Europe with its superior conventional forces without fear of nuclear retaliation (which is current NATO policy).	It would be far better to keep a war in Europe conventional in nature, if war were to occur. If the SDI forced the USSR to rely on conventional weapons, that would be a great gain, since such weapons are less suited for a surprise attack and can be recalled.
The SDI exploits the biggest advantage the United States has over the USSR: technology.	The USSR since World War II has matched US military technology in many areas such as atomic weapons, multiply-targetable reentry vehicles, and satellites. Technology could also be developed to defeat the SDI system.	Such a high-tech race is favorable to the United States. If the USSR is copying the technology, it will be behind the United States.

Table 9a: Points and Counterpoints in the SDI Debate: Arguments in Favour *cont.*

Policy arguments in favour of pursuing the Strategic Defense Initiative		
Point	Counterpoint	Rebuttal
The SDI will revitalize US high technology through spin-offs for both military and civilian applications.	The classified status of most SDI research will severely hamper commercialization. SDI research will also draw money and researchers away from commercial work. To pep up civilian industry, it would be more efficient to spend the money directly on civilian industry.	The jet engine and the nuclear power reactor are examples of technology that was commercialized from classified defense programs.
The SDI is a pure research program designed to explore whether a new doctrine of deterrence based on defense against nuclear attack rather than retaliation could and should be developed.	All SDI contracts — including those to universities — are under Defense Department category 6.3A, a restricted category for 'advanced technology development'. Many demonstrations of scaleable prototypes are planned.	The contracts were put in the 6.3A category primarily to make accounting easier.

Source: J. Adam and J. Horgan (1985), p.58

Table 9b: Points and Counterpoints in the SDI Debate: Arguments Against

Policy arguments opposed to pursuing the Strategic Defense Initiative		
Point	Counterpoint	Rebuttal
The SDI threatens to weaponize space, thereby escalating the arms race to new heights.	The purpose of the SDI is to end the nuclear arms race. It offers a chance to nearly eliminate nuclear missiles by treaty.	With the SDI, each side would be racing to improve its defenses, increase its offenses, and develop new counter-measures — a triple-headed arms race instead of the current single one.

The SDI will undercut the ABM treaty, which is based on the premise that the offensive strategic balance will be stabilized if defense is prohibited.	The ABM treaty bans no type of research; the SDI is a research program designed and executed within the ABM treaty. Also, the USSR is pursuing similar research.	Some of the demonstrations planned under the SDI research program could be perceived as 'components' in an ABM system, thus violating the treaty.
As a defense, the SDI will not be effective in protecting populations. Even if it were 95 per cent effective, some 450 strategic nuclear weapons would land in the United States in the event of a nuclear attack, doing incalculable damage to society. Actual estimates of its effectiveness are far lower.	Without the SDI, 10,000 weapons could land in the United States, negating any chance for a retaliatory strike.	Poseidon submarines now give the United States an invulnerable capability for massive retaliation. The SDI merely replaces one kind of deterrence with another kind of deterrence — the opposite of the way the President and the Secretary of Defense are trying to sell the program. Moreover, after being hit by 450 nuclear weapons, the United States would cease to exist as a nation.
Defensive satellites will always be more vulnerable than the missiles they are intended to destroy, this vulnerability leads to instability. In addition, the USSR has said that it will build more ICBMs to overwhelm any US anti-ballistic missile system.	The object of the SDI is to make it so costly for the USSR to overwhelm the system that the Soviets are discouraged from that approach. If it costs more to build offensive weapons than defensive weapons, then offensive weapons will be less valuable, making arms reductions more likely.	Any defensive system could be overcome by dozens of different types of countermeasures that confuse, overwhelm, underfly, or destroy the system. Already both the US and the USSR are investigating methods to defeat any space-based defense system. By the time SDI is deployed, the USSR would have already deployed a new generation of offensive missiles incorporating counter-measures.

Table 9a: Points and Counterpoints in the SDI Debate: Arguments Against *cont.*

Policy arguments opposed to pursuing the Strategic Defense Initiative		
Point	Counterpoint	Rebuttal
With the SDI point of view being opposite to that of the ABM treaty, the United States will have to persuade the USSR that defensive deterrence is superior to mutually assured destruction.	The USSR has always valued defenses, the SDI simply indicates that both sides are now integrating defenses into their programs.	While the USSR may spend vast sums on air defense, the Soviets also value the ABM treaty and have repeatedly said so. Talking them out of it may be difficult.
US allies worry that the SDI will accelerate the the arms race between the superpowers, endangering the rest of the world.	US allies are attracted by the possibility that participation in the SDI will spur their own high-technology industries.	US exports controls will severely limit any technology transfer out of the United States.
The SDI is prohibitively expensive and may severely tax the US GNP. Estimates range from $200 billion to $2 trillion for an initial system, which will constantly need updating.	It is not yet clear what the SDI will cost. Part of the program will investigate ways to hold down costs. The United States has been neglecting strategic defense in recent years. The program can be halted if proven unfeasible.	The cost of $26 billion by 1990 makes the SDI the largest research program ever projected. Research has historically occupied a fairly small percent of an entire RDT & E program for weapons systems. A program this large will be self-perpetuating whether or not it is feasible.
The SDI will be a pork barrel for defense contractors, with contracts spread through every congressional district.	There is nothing wrong in distributing government contracts as evenly as possible. Competitive contracts are being awarded to promote efficiency.	The reason that contracts are evenly spread among congressional districts is not to achieve efficiency but to ensure political support by seeing that the SDI means jobs in all districts.

Experts differ widely — and sometimes heatedly — on how the Strategic Defense Initiative will affect domestic and foreign policy. No one, however, denies that the effects will be profound. Above are summarized a few of the most frequently advanced arguments for favouring or opposing the SDI. These are merely assertions — *Spectrum* makes no claim as to the factual basis.

Source: J. Adam and J. Horgan (1985), p.59

government-capital-science social constituency is already shaping up behind SDI as the Departments of Defense and Energy are providing special funds to 'stimulate the widest involvement of technical talent within government, industry and universities' (Yonas, 1985, p.32). As a result, some large corporations such as Lockheed, GE and RCA have already formed special SDI divisions or named vice presidents to preside over SDI work, while the three national weapons laboratories — Los Alamos, Sandia and Lawrence Livermore — have accelerated their research (Adam and Fischetti, 1985). Despite all this, SDI is still at an early stage of development. Not only does it lack widespread rooting within the fabric of US society but, most significantly, it suffers from serious problems of credibility as the fantastic promises of total protection against nuclear attack have given way to much humbler objectives (Clery, 1987). In this respect, the most serious blow to SDI's scientific and political credibility was dealt in 1987 by a report from a panel of seventeen physicists from the American Physical Society, including three Nobel Prize-winners, who concluded that the state of knowledge in directed-energy weapons, one of the critical areas of SDI, was still too low even to determine the feasibility of the proposed SDI system. The study says

> Although substantial progress has been made in many technologies of DEW [directed-energy weapons] over the last two decades, the Study Group finds significant gaps in the scientific and engineering understanding of many issues associated with the development of these technologies . . . We estimate that even in the best of circumstances, a decade or more of intensive research will be required to provide the technical knowledge needed for an informed decision about the potential effectiveness and survivability of directed-energy weapons systems. (Patel and Bloemberger, 1987, pp.31-2).

This means that the initiative has yet to establish itself as a solid, long-term basis for the renewed drive of military interests. This is thus a difficult period for the military-led social constituency. It seems likely that a strong challenge to SDI may yet prove successful in thwarting its ambitious plans. Indeed, it is difficult to envisage a better time to make this challenge than the present, when sociotechnical commitments and hence the momentum of the SDI system are still relatively weak. This vulnerability, however, may not last long, since strenuous efforts are being made to build up the momentum of the military-led social constituency. In 1985, for instance, Brown assessed the situation as follows:

> At the moment little exists in the Star Wars program except some limited research and development activities . . . But I suspect that what we are seeing is the tip of the iceberg. The scientific-military-industrial community has sniffed the air, has smelled the billions of dollars which are potentially available, and has started to plan programs and write project proposals (Brown, 1985, p.3).

A year later the attempt to root SDI in US society was making noticeable progress. Plans to set up an SDI Institute near Washington DC were announced and, most significantly, a major campaign aimed at strengthening and expanding the SDI social constituency at grassroot levels was launched by the official Strategic Defense Initiative Organization (SDIO). This crucial development is described by Vandercook.

> The Strategic Defence Initiative Organization (SDIO), in response to requests from local businesses and politicians, is moving to establish the resilient base of local support that has proved so important to the irrepressible B-1 bomber (major contracts in 13 states) and the unstoppable MX missile (major contracts in 14 states). An example was the 'SDI Illinois Symposium' held last May 30 at Argonne National Laboratory outside Chicago that brought local industry and research executives together with much of the SDIO leadership from Washington . . . Local executives hoped to increase the amount of military research and development (R&D) money coming to the state and to their own companies. The SDIO's objective was to develop the local backing that might guarantee its budgets with Congress (Vandercook, 1986, p.16).

The significance of these actions has been well-expressed by the same author, in line with my own theoretical approach.

> The SDI program wants to develop and nourish a local constituency — one that will in turn nourish and protect it. With billions in funds to attract this local constituency, it will grow (*ibid.* p.18).

e) Conflicting Views About the Wisdom of the Renewed Military Drive

A military-driven social constituency looks set to have a long-term impact on the development of microtechnology in the US. However, unlike the 1950s, everything is not as consistent as the government-military social constituents would like it to be. Indeed, for many, with strong international economic competition, the renewed military drive has caused great concern. As MacKenzie explains,

> One great American fear is that while engrossed in its military role as a superpower, economic and technological leadership may quietly slip away to Japan . . . Precisely because military and civil technologies have to a degree diverged, a military obsessed America might be a poor commercial rival to a competition-oriented Japan (MacKenzie, 1985, pp.10-11).

Since this process is in its early stages, the impact of the military on the US microtechnological strength is not at all clear, and there are conflicting views about what is taking place. The semiconductor industry in particular has become the focus of much of the argument, chiefly because of its strategic importance[18] and increasing Japanese presence, although the Fifth Generation Computer follows a very similar pattern.

The case against strong military involvement in semiconductors stresses the difference between military and commercial development, and rests primarily on the fact that many of the microcircuits the military are interested in have relatively few civilian applications [*Aviation Week & Space Technology* (1981), Ferguson (1983), Sumney (1980)]. This has led some semiconductor firms to express concern that the VHSIC program will divert scarce resources from civilian projects, especially research and development, thus holding back commercial technological progress [Botkin *et al.* (1982), De Grasse (1984), Mowery (1983)].[19] Likewise, in relation to the proposed Sematech consortium,

> many observers see a Defense Department-backed manufacturing consortium as a mixed blessing. The fear is that Pentagon financial support would mean Pentagon control,

> diverting the consortium from its announced purpose of support for commercial manufacturing (*Electronics,* 22 January 1987, p.29).

The crucial point in both cases is that, as Borrus *et al.* (1982) point out, competitive civilian development will not automatically emerge from defense expenditure.[20] Indeed a report on VHSIC gave much support to these apprehensions by suggesting that

> Contrary to the expectations of many companies participating in the Very High Speed Integrated Circuits program, speeding the pace of semiconductor efforts and, hence, commercial spinoffs, are apparently not central to the Defense Department project . . . export and national-security restrictions, a structural bias against innovative smaller firms, and a military-systems requirement working at cross purposes with merchants' needs will severely limit the commercial benefits derived from the $680 million program (*Electronics Week,* 17 December, 1984, p.60).

The fear has been that military-funded programmes like VHSIC would handicap US firms competing with the Japanese for the leadership in VLSI technology (Levin, 1982a). And if one considers that the Japanese have been making important advances in semiconductor technology, and have already overtaken the US in key areas of VLSI, these fears are justified. Figure 18 illustrates how a five-year lag over the transistor in 1948 had been transformed into a one-year lead with 256K RAM memory in 1982. Japan eventually captured ninety percent of the 256K RAM market (Wallich, 1986). In turn, Figure 19 gives an idea of the present trends affecting a wide range of US semiconductor technologies relative to Japan. From twenty-five categories shown, the US is leading in only four while Japan is leading in twelve. More significant for the future, in nineteen out of twenty-five categories the US position is considered to be deteriorating, with the other six categories remaining equal. Figure 18 also illustrates that Japanese success has been the result of long-term government programmes spearheaded by the Ministry of Trade and Industry (MITI) and aimed at developing the country's semiconductor industry towards commercial markets and international competition.[21] Below I shall deal with the Japanese electronics industry in detail, but it is worth stressing here that the Japanese social constituency of microelectronics technology has no appreciable military component and, as a result, has consistently concentrated its efforts on the commercial development of the tech-

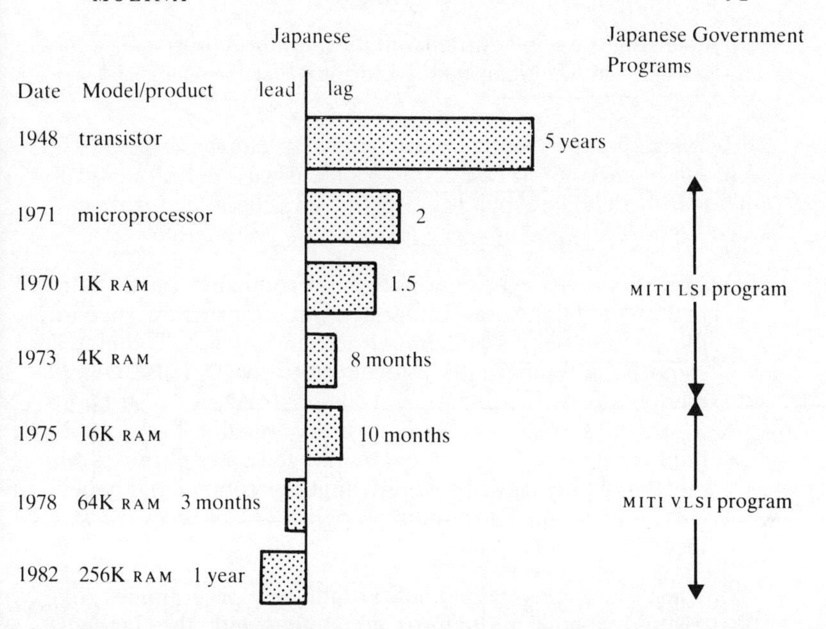

Figure 18. Evolution of Japanese Position Regarding the Frontier of Various IC Products (1948-82)
Source. C. Freeman (1985), p.220

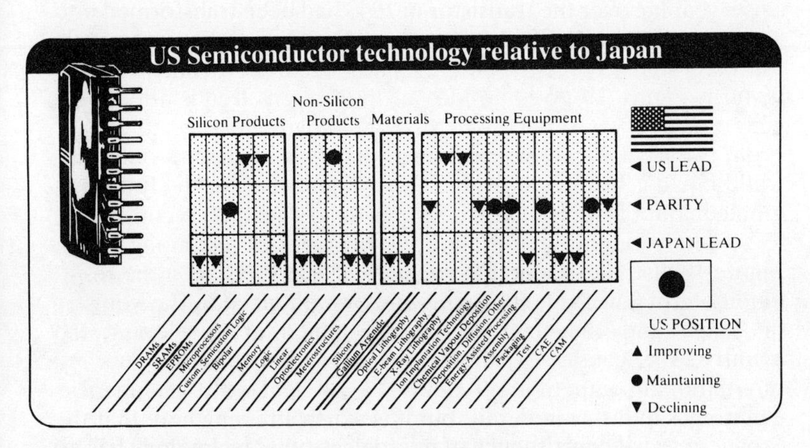

Figure 19. US Semiconductor Technology Relative to Japan
Source. Financial Times, 27 January 1987, p.15

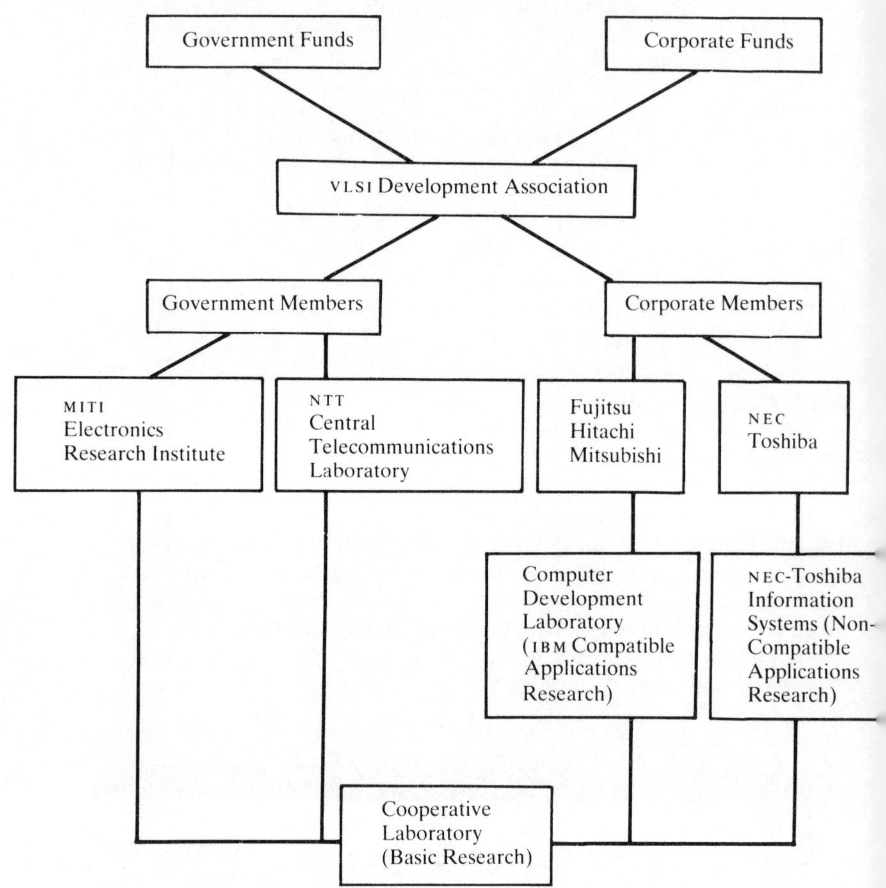

Figure 20. Government-Capital-Science Constituents Behind the Japanese VLSI Project
Source. UNIDO RCSB (1985), p.18

nology.[22] Figure 20 shows the organisation of Japan's VLSI project and the way in which the specific government, corporate capital and science constituents have interlinked behind the development of the VLSI technology. It is in this context that the US divergence of resources towards military ends is seen as harmful. Thus, the

> potential commercial spin-off of the program [VHSIC], primarily in the high-resolution technology, is not thought to be worth the effort. Such technologies are being developed

> by industry anyway, and a better way of spending money would be in research on VLSI system design (Hazewindus, 1982, p.145).

Indeed, after some years of the VHSIC programme running, a 1984 study claims that the commercial sector is in fact keeping pace with the military as much in miniaturisation as in computer-aided design and engineering (*Electronics Week*, 17 December 1984). As a more fruitful alternative for commercial purposes, it has been suggested that a major programme similar to VHSIC might be supported by the government, based on civilian pursuits which would directly benefit the US semiconductor industry (Borrus *et al.*, 1982). However, under the present circumstances this is unlikely to happen, at least as long as the pressures of war bring the military the renewed leverage it now possesses within the social constituency of microtechnology. Recent developments concerning the proposed Sematech consortium seem to confirm this assessment.

In marked contrast to this position, the VHSIC programme and the re-emergence of the military as a major force in the development of microelectronics has been welcomed by some sectors as an important contribution towards strengthening the US semiconductor industry. Indeed, given the participation of industry and university shown in Tables 7 and 8, many companies appear to believe this, particularly large corporate capital leading the teams' work. As one report states,

> Close alignment of the semiconductor industry's technology goals with those of the defense community for the first time in more than a decade has generated strong support among the components suppliers for the Defense Department (VHSIC) program (*Aviation Week & Space Technology*, 1981, p.77).

The reasons advanced are threefold. First, the military is providing important development funds; secondly, the military market looks set to grow in importance as the VHSIC programme aims for a class of semiconductor products that will be in high demand throughout the defense-systems realm; and, thirdly, VHSIC products will be based largely on the same technologies that the semiconductor companies are already exploring [(*ibid.*), *Electronics* (28 September 1978)]. The Sematech project is creating the same kind of arguments and, being aimed at manufacturing processes rather than specific products, the case for convergence of military and industrial interests seems to be stronger. As one report has put it,

> The goals of the Pentagon and the US chipmakers are different. The Pentagon wants a secure American-based supply of advanced chips while the industry aims to share the huge costs of developing next generation production technology that will enable it to continue to compete with Japan. In proposals for an industry-wide co-operative manufacturing project, both the military and the industry appear, however, to have found common cause (*Financial Times*, 27 January 1987, p.15).

In other words, what is in conflict with the current needs of the industry for some analysts is a perfectly complementary development for supporters of military participation. As Levin (1982a) has argued in the context of VHSIC, the 'fears seem to have been misplaced, as industry participants have come to recognise the substantial complementarity between VHSIC and commercial VLSI objectives' (Levin, p.50).[23] Nevertheless, VHSIC has not helped alleviate the decline of the US semiconductor industry in world markets. Unlike the case of SCP, some argue VHSIC was not conceived as a direct response to the Japanese VLSI programme. Currently, this would be Sematech's purpose, implying a recognition of VHSIC's negligible impact on the competitiveness of the US semiconductor industry. Sematech, however, is just beginning its development; in direct response to the Japanese challenge, the main approach of US firms so far has been to seek more direct assistance from the government. For instance, the 1981 legislative program of the Semiconductor Industry Association (SIA) has three major components: tax incentives for R&D expenditures; access to the Japanese domestic market by relaxing tariffs and controls on direct investment; and support for engineering education [(*ibid.*), Borrus *et al.* (1982)]. In this context, VHSIC is seen as merely providing an indirect stimulus to US companies competing in world markets, insofar as it has stimulated these companies' technical programmes. In fact, as VHSIC got under way, reports suggested that 'there is widespread agreement among companies involved in the program that it has accelerated their previous corporate timetables by at least two or three years' (*Aviation Week & Space Technology*, 1981, p.52). See also Iversen (1985).

So far the case against the military is making little headway. On one hand, there is thought to be commercial spillover and, on the other, the absence of any alternative civilian programme of similar scale makes it impossible to make comparisons over VHSIC's relative benefits. As a result, it seems clear that the military are well-placed to play an influential role in microelectronics develop-

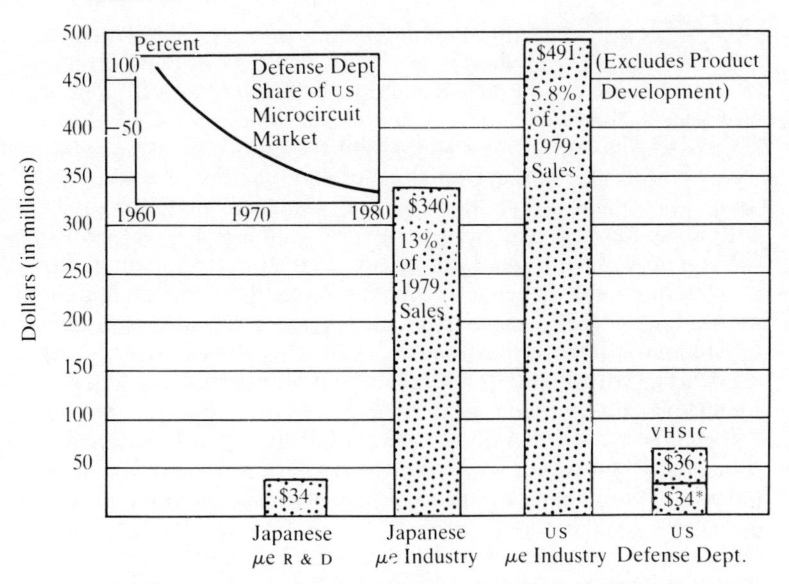

* Microelectronics Research and Exploratory Development

Figure 21. Annual R & D Support for Microelectronics in the US and Japan. Year 1979

Source. Aviation Week & Space Technology (1981), p.53

ment for some time. Clearly, Sematech is already showing similar potential as well as indicating the way in which such an influence is likely to make itself felt in the future.

The real test for the military-backed US semiconductor industry, however, will be its performance in international competition and, particularly, its ability to counteract the challenge of the Japanese social constituency. In this respect, it is relevant to examine the volume of VHSIC funds for R&D in microelectronics in comparison with the volume of US industry's own effort and the volume of Japan's effort. Figure 21 shows the situation about the start of VHSIC, in 1979. It can be seen that funds from the DoD were less than fifteen percent of the funds provided by the US industry for R&D in microelectronics. However, since VHSIC aimed at securing a similar amount from industry, the funds devoted to the programme become approximately one-third of the industry's support for microelectronics R&D. In contrast, the Japanese devoted fewer R&D funds in absolute terms, but not only are these funds concentrated on the commercial market, they also represent a much larger proportion of the industry's sales, 13% as compared with 5.8% for the US industry, thus showing the crucial importance the Japanese attach to microelectronics R&D in the battle for the industry's leadership.

Most significantly, Japanese R&D investments take place within a very different context from that of the US. For instance, the Japanese possess an ability to work out efficient and reliable methods of mass production; an unparallelled willingness to invest within a more favourable internal financial climate; and an ability to cooperate, pooling resources from both the government and the private sector. Also, R&D costs in Japan are generally lower and, until recently, Japanese companies had been involved in a 'catching up' strategy which has been less costly in R&D terms.

Cultural traits have also been highlighted, particularly those seen in management and labour practices and in a rapid adaptability to change [Sigurdson (1983), *Science* (1986), Kikuchi (1983)]. It is from this background that Japanese investments in electronics R&D must be compared with the United States' research and development investments and their major military concern. The result of these processes cannot be known for some time. In the meanwhile, we can only note that according to Figure 18, since 1978, the year of the announcement of VHSIC, to 1982, the Japanese lead in RAM memory had increased from three months to one year[24] and, as will be seen later, it is estimated that Japan may become the dominant force in the future semiconductor market.

The impact of the renewed leverage of the military within the social constituency of microtechnology also gives rise to some issues which, while related to the economic battle for the leadership in the market place, are of wider social implication. In this respect, serious concern has been expressed about the social impact of the military attempt at directing the development of microtechnology for its own purposes. One of the main worries is over the military's tendency to impose security regulations on the R&D work carried out by industry and universities. For instance, foreign researchers are expected to be excluded from the projects of military interests and circulation of information will be restricted [Hazewindus(1982), Dickson (1984)]. This has caused some friction among the social constituents, since such measures blatantly contradict normal academic practice and may inhibit industrial innovation and competitiveness.

Originally, the issue centred around the university in relation to research in cryptography but a cursory look at the DoD's 'list of critical technologies whose acquisition by potential adversaries would be detrimental to national security' (*Federal Register*, 1980, p.65014) demonstrates that the military considers the whole of microtechnology an area of national security. Of seventeen entries in the list fourteen relate to microtechnology. Such military interest in microtechnology and the attempt to enforce controls upon its diffusion has meant, for instance, that Cornell University has withdrawn from VHSIC following secrecy over the third phase of the programme, which was seen as damaging for university research in microelectronics (Botkin *et al.*, 1982). For similar reasons, the Department of Commerce has clashed with the DoD over the latter's control on the export of about 10,000 products 'simply because they have embedded microprocessors' (*Electronics News*, 7 March 1983, p.6). It is argued that many of these products are freely available in the world market so that unilateral export controls on them can only harm the competitiveness of US exporters (*ibid.*). This is confirmed by a study from the National Academy of Sciences which estimates that in 1985 the American industry as a whole lost $9.3 billion of orders and 188,000 jobs (*The Economist,* 6 February 1988).This concern has recently prompted the US electronics industry with the support of the Commerce and State Departments to propose revisions to 'export control regulations for microcomputers and related equipment that would dramatically liberalize high technology trade with the Eastern bloc' (*Financial Times,* 18 August 1987, p.26). These moves have helped critics of the military spin-off argument greatly to reinforce their case. Robert Noyce, for instance, has pointed out that VHSIC 'was originally justified on the basis that it would have commercial potential and benefit. But right now the Defense Department is trying to classify the whole thing. The military tends to go its own path instead of reinforcing the commercial activities' (quoted by Adam and Horgan, 1985, p.62). The same problem exists for the commercial potential of SDI, since the rules restricting exports of high technology are bound to include SDI technology. Indeed, in a clear show of military appetite for control, 'These rules have been expanded recently to include not only weapons but also fundamental technologies with no proven military application' (*ibid.* p.63).

Another major concern expressed by some analysts is the impact of ManTech on the United States labour structure. All ManTech projects are aimed at reducing the need for skilled labour and are bound to result in a net loss of jobs (Tirman, 1984). Yet the military has ignored this issue since, in its search for control and automation, it has given labour virtually no say over the philosophy or design of manufacturing technology programmes, [Noble (1984), De Grasse (1984)]. How many jobs? and what will they be like?

Are the technologies really more productive? If so, productivity for whom? asks Schlesinger (1984). No answers to these questions have come from the military although they seek credit for leading the reindustrialisation of the US.

From a political angle, military programmes, in particular SDI, have been criticised for contributing to the arms race [Brown (1985), Panofsky (1985), von Hippel (1985)]. In addition, Tucker (1985) fears that SCP will enhance the possibilities of US military intervention in the Third World as robots replace soldiers and human casualties appear more acceptable.

It seems plausible to conclude that the military effort to gain control of microtechnology for its own purposes has, at the same time, wider social, economic, political and cultural implications at both national and international levels. Yet in pursuit of its own interests, this responsibility does not seem to deter the postwar social complex from pushing forward the above developments.

CHAPTER FIVE

THE CURRENT DEVELOPMENT OF MICROTECHNOLOGY II

Interacting with the relations and trends seen in the previous chapter, there are two other processes which dominate current development of microtechnology.

Convergence of Technologies and Capital in the Electronics Industry

Convergence is one of the key features of the microelectronics revolution. Its importance is seen particularly in industrial convergence which has itself developed out of the growing convergence of electronics systems on the one hand, and the expansion of large corporate capital from other industrial sectors into the electronics industry on the other. Two levels of industrial convergence can be related to the convergence of electronics systems.

i) the convergence between producers within the electronics industry itself, and

ii) the convergence between industries directly applying electronics technologies or most likely to be affected by their expansion.

Convergence within the electronics industry may be said to have closely followed that of technology. Thus, when the microprocessor transformed computer power into an electronic component, what had hitherto been the province solely of the computer industry inevitably began to overlap, at least at the lower end which was occupied by microcomputer and minicomputer systems. Robinson has described the evolution of component manufacturers since the days of the transistor.

> In the days of discrete transistors, components manufacturers made transistors, while someone else designed circuits for specific uses. As integrated circuits became more complex,

> components manufacturers became circuit designers as well as part makers . . . But today's most sophisticated microprocessors are as powerful as minicomputers, and chips yet more powerful are on the drawing boards . . . components manufacturers are now makers of computer systems and must provide the same kind of software support long characteristic of large computer companies (Robinson, 1980, p.483).

Thus, semiconductor processing, systems design and software development have converged into the semiconductor industry, reshaping the industry in the process. Following increased need for resources[1] and, consequently, rising costs alongside growing Japanese competition, US semiconductor companies were compelled to expand into the equipment and end-product market, mainly computers and consumer products [Hazewindus (1982), Mowery (1983)]. In the latter market, for instance, companies such as Texas Instruments and National Semiconductor moved into calculators and watches realising that it was more profitable to use their own chips for making the whole products. Most importantly, however, semiconductor companies began to invade computer territory as they 'tried to evolve from just selling bare components to supplying complete systems of chips mounted on a circuit board and partly programmed for a particular application' (*Financial Times,* 12 September 1980, p.11). Thus, in 1981, Intel introduced a microcomputer on a board (the 4-3-2), claiming it had the power of a small mainframe. In turn, National Semiconductor began to supply Japanese designed computers (*Financial Times,* 1 July 1981, p.17), while TI's considerable experience in the computer field[2] influenced the design of its microprocessors (*The Economist,* 8 November 1980). By the late 1970s, there were reports of legal battles between minicomputer and semiconductor companies for alleged invasion of each other's fields. Fairchild and National Semiconductor, for instance, had been sued by Data General and DEC for imitating their machines (*Business Week*, 19 March 1979).

As semiconductor companies expanded into equipment, however, computer manufacturers began to expand into components both to counteract the onslaught of semiconductor companies by producing chips closer to their needs and, more particularly, to reduce their dependence upon these companies. In so doing they also avoided the risk of bottlenecks or of being caught in a shortage of chips which would seriously jeopardise their markets (Lamborghini, 1982).[3] Thus, by the late 1970s, the three largest minicomputer companies, Data General, DEC and Hewlett

Packard had all begun to invest heavily in the upgrading of their semiconductor capabilities (*Business Week*, 19 March 1979). On the other hand, IBM, just like AT & T (the Bell System), has always produced integrated circuits for its internal consumption, although until recently it was prevented by regulation from entering the semiconductor market.[4] Of other large computer companies, Honeywell has also made big investments in semiconductor capabilities (Ferguson, 1983) while NCR, Burroughs and Sperry all have production of integrated circuits for internal consumption only, i.e. captive production of integrated circuits (Hazewindus, 1982). The convergence of technologies has thus been largely responsible for the vertical integration between semiconductor and computer industries.

a) AT & T and IBM on a Collision Course

Nowhere has industrial convergence been more dramatically illustrated than in the convergence of the computer and the telecommunications industries, epitomised in the battle of the giants AT&T and IBM. As one commentator has pointed out,

> For the first time AT&T and IBM, the undisputed champions in their respective classes, are set to compete in the same league. The previously rigid barriers between the telecommunications and computer business have been eroded by technology and by major changes in US policy, leaving the two giants staring eyeball to eyeball (*Financial Times*, 18 January, 1983, p.22).

Although AT&T and IBM have competed in the past to a certain extent,[5] the regulatory controls around telecommunications (consent decree 1956) had largely insulated this field from market competition, leaving AT&T as the dominant force while preventing it from openly competing in other fields such as computers. In turn, the computer and electronic data processing (EDP) fields were unregulated, allowing competition, but still dominated by IBM. In terms of market and legislation, therefore, for years everything seemed clearcut,

> IBM sold computers and AT&T communications services, and each became a top-ranking customer of the other. They fought their battles and defended their titles in separate games played on different boards — IBM in a competitive world, AT&T in a regulated world (Johnson, 1982, p.25).

In 1982 this 'happy' situation officially ended as a settlement on deregulation in the communications field was finally agreed. The distinction between voice and data transmission on international routes has been abolished, enabling AT&T to offer computerised data services abroad while permitting other companies to enter the international telephone services (*Financial Times,* 10 January 1983). Indeed, AT&T is now free to enter almost any business it chooses, nationally and internationally,[6] it can sell as well as lease equipment and branch out into businesses like data communication, electronic information processing and office automation (*ibid.* p.12). AT&T is moving exactly in that direction. In so doing, however, the company has embarked on a course which may have great implications for the telecommunications industry, as well as for such industries as computers, consumer electronics and publishing (*Financial Times*, 18 January 1982). At the same time, AT&T's own territory is now open to competition. Indeed, electronics has now become a potential arena for the competing interests of electronic corporate capital, particularly AT&T and IBM.

Various commentators have looked at the strength, strategies and first moves of both AT&T and IBM,[7] and both giants seem to be well-matched. For instance, after the break up, AT&T's assets amount to $39 billion while IBM's assets are $38 billion (Bylinsky, 1984). Both companies spend about $2 billion a year on research and development and command a broad range of technologies. Bell Labs, however, is stronger on basic research and IBM is better oriented towards the development of broad product lines [*Financial Times* (18 February 1983), Whitington (1982)]. As a money-raiser, AT&T is in the stronger position but IBM has the stronger marketing organisation with an integrated worldwide system unmatched by any other company (Whitington, 1982). In terms of strategy, both companies emphasise that they are not aiming for a head-on collision, but AT&T has made it clear that it expects to emerge 'near the top in computers in the late 1980s' (Bylinsky, 1984, p.50).[8]

AT&T counts on its UNIX operating system (software) as the main lever in its plan to create a huge new computer business (*ibid.*) and is building a nationwide system, the Avanced Communications System (ACS), which will 'translate' between different makes of computers.[9] At the same time, it is planning to sell a range of small computers and terminals which can handle both data and voice communication. In pursuing this strategy, however, AT&T aims 'not to be just another computer company but to use computers to complement the other communications services' it offers (Kozma, 1985, p.27). This approach seems to have won AT&T a major contract from the National Security Agency for supplying a network of 325 UNIX-based 32-bit minicomputers. On the basis of this contract which is said to be well over $1 billion, AT&T has already joined the ranks of computer vendors in the US (*ibid.*). More recently, the company has also launched AT&T Mail, a commercial electronics mail service which allows computer and terminal work station users to send messages across telephone lines to other AT&T mail customers (*Financial Times,* 26 February 1986). In turn, IBM is also developing integrated voice and data work stations and has moved into telecommunications through the acquisition in 1984 of Rolm, the third largest supplier of private exchanges (PBX's), and through its 16% equity investment in MCI Communications Corp., seen as 'an offensive strategy to strengthen the only viable competitor to AT&T's communications business' (Kozma, 1985, p.29).[10] In addition, IBM has begun to use UNIX in an attempt to influence its development within the computer field (*ibid.*) In pursuing these strategies, both companies are moving towards office automation, with AT&T planning to use its ACS as the core of such a development [Wohl (1982), *Financial Times,* 18 January 1982]. IBM has recently launched a personal computer-based system for managing data and voice (*Financial Times,* 2 February 1987) and also has big stakes in factory and banking automation having recently become the leader of the CAD/CAM market (*Financial Times*, 16 July 1985) while being second in the league of automated telling machine manufacturers (Bessant, 1984).

It is far too early to say what will happen in the confrontation between these two corporations. The challenges ahead are exacting and, for all their strength, it will not be easy for IBM and AT&T to realise their ambitious goals. Instead, in a competitive world of rapid technological change, their development will probably be slow, a process of trial-and-error. Already, there is evidence to support this view. A recent report for example states that 'despite costly efforts to diversify, it has proved as difficult so far for IBM to make a profitable business out of telecommunications as for AT&T to master the art of selling computers' (*Financial Times*, 2 February 1987, p.11). In particular, IBM has had problems with its Rolm venture, estimated to have cost $100 million last year, while a joint venture in electronic financial information with Merrill-Lynch was scrapped after poor results (*ibid.*). On the rival side, AT&T last year abandoned its net 100 data transmission system (a business-oriented computer communications network) after investing about $1 billion over ten years. The system is said to have been made obsolete by the development of much more powerful desktop and minicomputers. In addition, AT&T's venture with United Technologies to provide information services to tenants in individual buildings also produced poor results, and the company withdrew (*Financial Times*, 23 January 1986). These developments have meant that AT&T has now shifted its strategy more emphatically towards systems integration — making machines talk to each other — rather than computer hardware itself, a move that plays to its strength in transmission (*Financial Times*, 15 April 1987).

Given this rather confused scenario, it seems that the future remains as vague as Fishman suggests: 'The battle for control remains. The only certainty about its outcome is its uncertainty: despite the wealth of forecast, it is difficult to know what the market will look like in five, ten or twenty years' (Fishman, 1981, p.318). Only one thing seems to be clearer, namely, that given the convergence of electronic systems, in the long-run, the battles currently gathering momentum are only the first manifestations of 'a struggle between the giant electronic corporations for control of the "information structure"' (Webster and Robin, 1979, p.297). And in this battle, IBM and AT&T, although the most powerful corporations, will not be the sole contenders. As the technologies converge, other companies are already joining the fray, including industrial interests previously unrelated to electronics, now drawn into the field either to protect their businesses or to gain a stake in the growing electronic market which is expected to reach approximately $520 billion by 1990 (*Electronics & Power*, September 1987).

b) Convergence of Capital

Within the electronics field, companies other than IBM and AT&T are aiming to secure a place in the electronics markets. In data transmission, for instance, Xerox, ITT and four value-added common carriers[11] (i.e. companies which provide services through the telephone network), themselves part of companies offering telecommunications and computer services, are already in competition. In the PBX market, participants include telecommunications companies such as US's GTE and Rolm (IBM) and Canada's largest telecommunications manufacturer Northern Telecom and Mitel a PBX manufacturer. Computer companies, Honeywell and Datapoint have entered this field too. The market for satellite communications is another area in which important electronics companies are competing, particularly RCA (now acquired by GE), ITT (now merged with Alcatel-France), GTE and Fairchild. The latter together with Continental Telephone, a major independent telephone company operating in the US and Canada, owns American Satellite Corporation which is, in turn, linked with Western Union's

Westar Satellite System (*Financial Times*, 18 January 1982). In the electronics mail market, ITT and GTE are already contending with their computer mail-box service while MCI Communications, the long-distance telephone company, has recently moved into it with serious intentions (Louis, 1984). The office automation market has attracted such names as Xerox, Burroughs and Philips and, as we shall see, is one of the areas where non-electronics giants are also converging. In all, it seems clear that a number of companies are set to play their part alongside the main contenders, IBM and AT&T, in the struggle for control of the changing and newly developing markets in electronics.

As we have distinguished at the beginning of the discussion, the electronics-based convergence of industries is not limited to the electronics industry. As the converging technologies spread themselves throughout the technical base of society, those industries most likely to be affected by the expanding electronics systems have also been dragged into the electronics net in order to protect their businesses and/or simply to take advantage of the broad range of available opportunities. Among these industries the most conspicuous are the aerospace, the print and mail industries [de Sola Pool (1983), Brock (1981), Irwin and Johnson (1977)]. In aerospace, for instance, firms with satellite expertise have seen satellite communications as a natural extension of their operations. Thus, along with the electronics companies we have seen above, both Hughes Aircraft (now part of General Motors) and Lockheed have also entered this market (Brock, 1981). With the print industry, since electronics has removed the barriers protecting publishers and newspapers, electronics companies offering media services have been invading this field. AT&T, for instance, has been taken to court by a group of newspapers to prevent it from publishing electronic yellow pages (de Sola Pool, 1983), while the arrival of electronic mail is bringing the postal service into direct competition with communications companies. The US Postal Service (USPS) attempted to enter the market in 1978 through a contract with Western Union for joint provision of Electronic Computer Originated Mail (ECOM). But, after considerable controversy involving even the Federal Communications Commission (FCC) and the Postal Rate Commission (PRC), ECOM was eventually withdrawn from the market [Brock (1981), Louis (1984)].

Finally, as was also outlined earlier, there are a number of huge non-electronics corporations from fields such as oil, engineering and the car industry who are converging into electronics to satisfy the changing needs of their own technologies[12] and, most importantly, to gain a major stake in the growing electronics market, particularly in such demanding areas as office and factory automation. The most conspicuous cases are those of Exxon, the largest US industrial firm, GE, the largest US engineering firm, and, more recently, GM, the largest US car manufacturer. But other big corporations are in the race too. Thus AM International Inc. acquired about a dozen companies to enter information processing, Eastman Kodak and 3M are working in office automation and various oil companies — Sun, Standard Oil (Indiana) and Gulf — have moved into word processing [McCartney (1978), Louis (1984), *Business Week*, 16 July 1984].

The scale of Exxon's, GE's and GM's efforts, however, is enormous and has involved the acquisition of many electronics companies. Exxon, for instance, has entered into chips and microcomputers, communications and terminal products, graphics and other intelligent terminals, facsimile transmission equipment, semiconductor laser products for computers, and is also involved in speech recognition equipment for voice input and voice response for computer data entry, information retrieval and telephone network applications [Fisher *et al.* (1983), McCartney (1978)]. Exxon acquired about fifteen companies to form its Information Systems Group, a system which when effectively integrated may make the company one of the leading suppliers of electronic data-processing products and services. In turn, GE is already offering an electronic mail service (Louis, 1984) and is 'aiming to become the major US supplier of automated "factories of the future"' (Kaplinsky, 1984, p.3), a market being contested by Westinghouse and IBM, the current leader in CAD/CAM, and, more recently, by the computer company Sperry (*Financial Times*, 5 November 1985). As a result, GE has gained and/or improved its foothold in microelectronics, CAD/CAM, robotics, numerical controls and also in computer services where it aims to become the largest computer service firm in the world (Kaplinsky, 1984).

Finally, GM's big move into the electronics industry took place in 1984 when the company agreed to lay out $2.5 billion to buy Electronic Data Systems Corporation. The latter company will form the cornerstone of a GM information processing company together with an AI company and a quality consulting group where GM has acquired strategic stakes (*Business Week*, 16 July 1984). GM has also acquired Hughes Aircraft for $5 billion and is now at the head of the satellite-makers league [Cortes-Comeres (1986), Zorpette (1986)]. The recent acquisition of RCA by GE for $6.25 billion has attracted most attention, however. This combined GE-RCA organisation not only has made GE the seventh largest industrial corporation in the US with $38 billion in sales, but, according to one commentator, has put the company 'on the threshold of becoming a high-tech giant second in sales only to IBM' (Cortes-Comeres, 1986, p.65). See also *Electronics,* 6 January 1986.

The convergence of technologies and their spread throughout society has thus underlain the convergence of previously separate industrial interests. As a result, electronics has become the arena for a major struggle as corporate capital from within and outside the traditional electronics industry strive to gain control over its development. The outcome of this process will depend upon many factors and, as will be seen below, the struggle will be world-wide.

Electronics Competition on a Global Scale.

In all areas of electronics, technology has been making rapid advances. The synergistic effect of the technologies upon each other, the competitive thrust, particularly of the microelectronics industry, and the exacting demands of the military, have clearly been behind this development. In addition, as different industries have begun to converge, their territories are being blurred. Many companies are being forced to expand and diversify their technologies in order to protect, improve or gain positions in the rapidly growing electronics market. As a result, for most companies, their need for resources has increased markedly, while requiring a fast return on investment. This process in turn has led to increased barriers hindering entrance to the markets, with high R&D expenditures, large-scale economies and a world-wide market becoming critical elements in the electronics industry.

Take the case of the microelectronics industry, for instance. Here,

> Increasing IC component density has been made possible by more complex processing equipment. Process equipment has become much more expensive as a result of its increasing complexity. Thus, the capital or fixed costs of IC production have increased substantially . . . At the same time, most IC prices have declined almost as fast as costs per unit for the new high capacity plant. The consequence of these two trends is the necessity for IC companies to operate at very high volumes. More than ever, high volumes are required to amortise higher capital equipment costs. And if it is to generate products which are competitive with other high-volume producers, the firm must constantly upgrade its fabrication plant to produce

> the newer denser ICS . . . Historically, this problem has been compounded by rapid technical change which makes process equipment obsolete and hence further increases capital costs (Rosenberg, 1980, p.361).

This situation is easily confirmed. For instance, according to R. Noyce, competing in VLSI markets 'implies the willingness to risk at least $50 million to a $100 million a year in capital, plus another $50 million to a $100 million a year in R&D cost' (quoted by Ernst, 1981). When one considers that about $5 million was adequate to start a state-of-the-art semiconductor company in the 1960s (Robinson, 1980a), it is clear how dramatically the barriers to entry have multiplied in the IC industry. Now a wafer facility for more than $80 million sales volume is seen as a minimum requirement (Hazewindus, 1982), as is the corresponding volume of business. According to Ernst (1981), from the mid 1960s till the late 1970s, there was a twenty-fold increase in equipment costs for wafer fabrication. This sort of development has meant that while in 1970-5 approximately forty cents of investment was needed to obtain a dollar in new sales, in 1975-80 this increased to fifty-five cents and, currently, to as much as seventy cents (Hazewindus, 1982). The figures are equally impressive for R&D. In 1978, for instance, the R&D outlays as percentage of sales were 5.8 for the semiconductor industry while the average for the US industry as a whole was only 1.9%, and even for the high-technology industries the average was 4.0% (Ernst, 1981).[13] These features underlie Rada's point that 'A characteristic of information technology, whether at a component, computer and telecommunications level, is that it can only be fully and economically developed in a world market' (Rada, 1980, p.114).[14] This process has been compounded by the growing presence of Europe and Japan in the world and the US markets, most particularly the Japanese who have been successfully challenging the US electronics predominance, thus greatly intensifying competition on a global scale. As will be seen, the combined effect has been to stimulate the capitalist tendency to concentrate capital, at both national and international levels.

a) Evolution of United States', European and Japanese Electronics Markets.

The crucial role of markets for industrial and technological development is well-known, hence the relevance of looking at the trends characterising the evolution of world markets for major sectors of the electronics industry. The set of figures from Figure

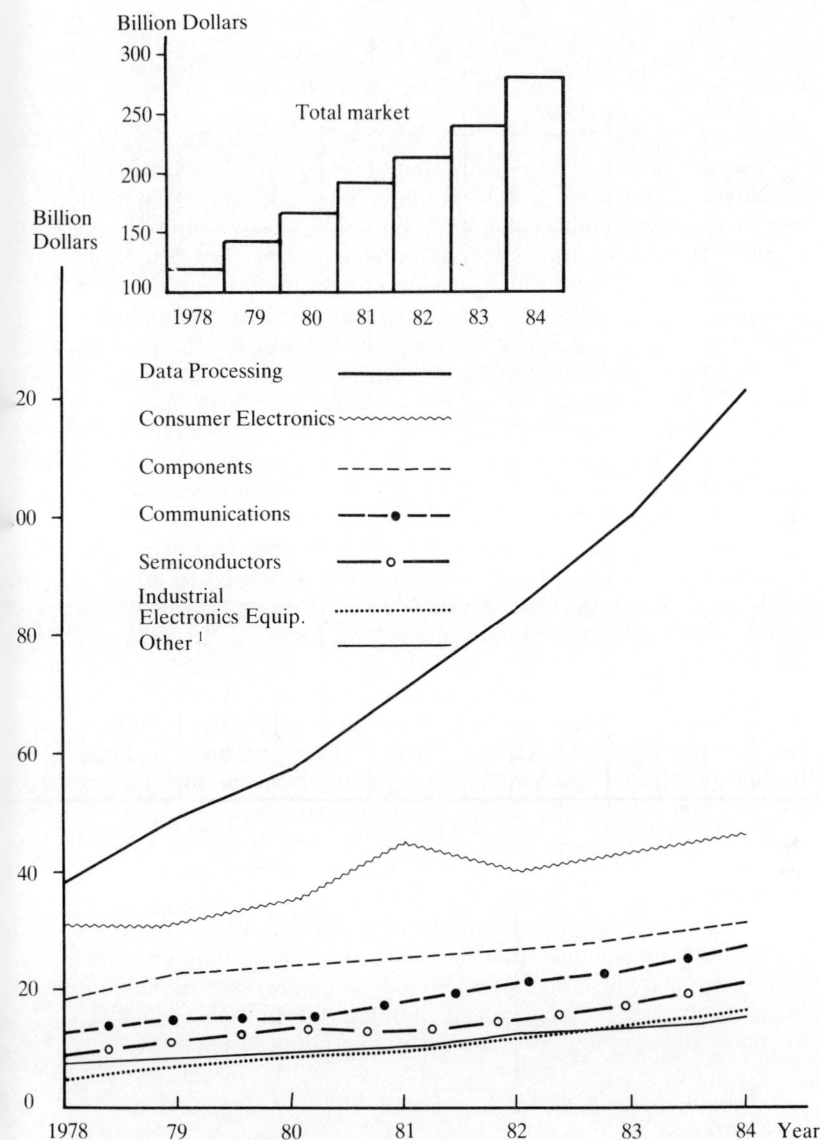

Figure 22. Total US, Japanese and European Market for Electronic Equipment and Components (1978-84)

[1] Test and Measuring Equipment, Medical Equipment

Source. Based on figures given in Molina (1987), p.258

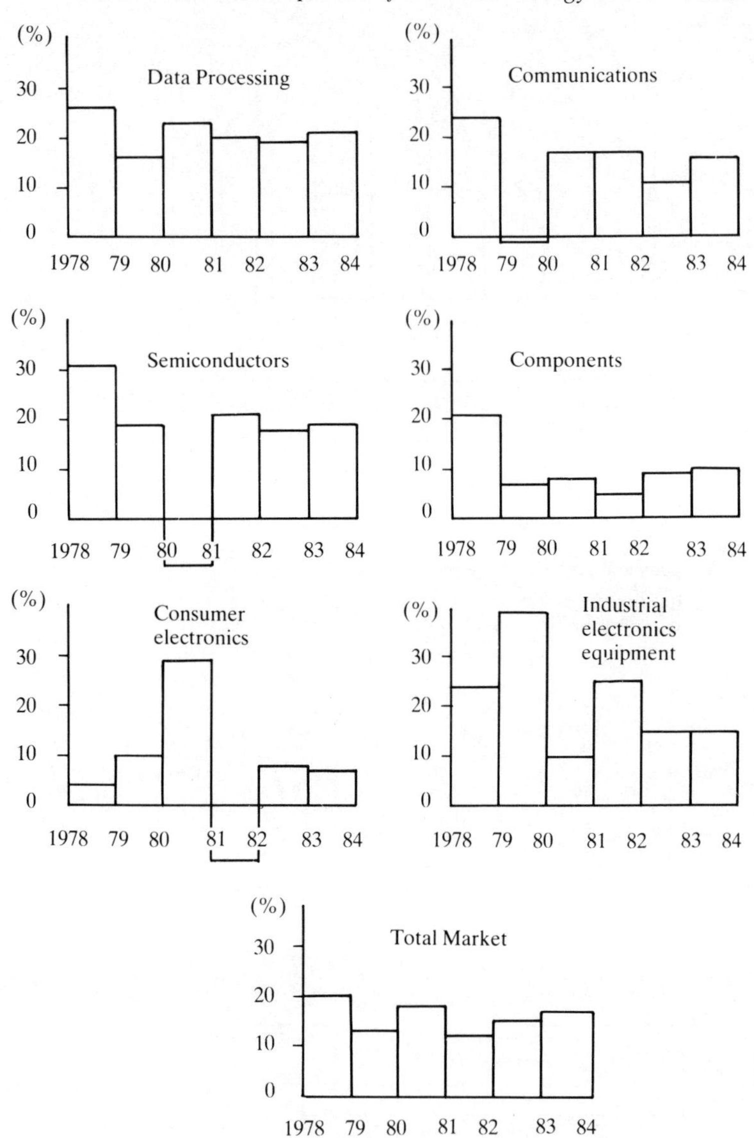

Figure 23. Annual Growth Rates for the Total US, European and Japanese Market for Electronic Equipment and Components (1978-84)

Source. Based on figures given in Molina (1987), p.259

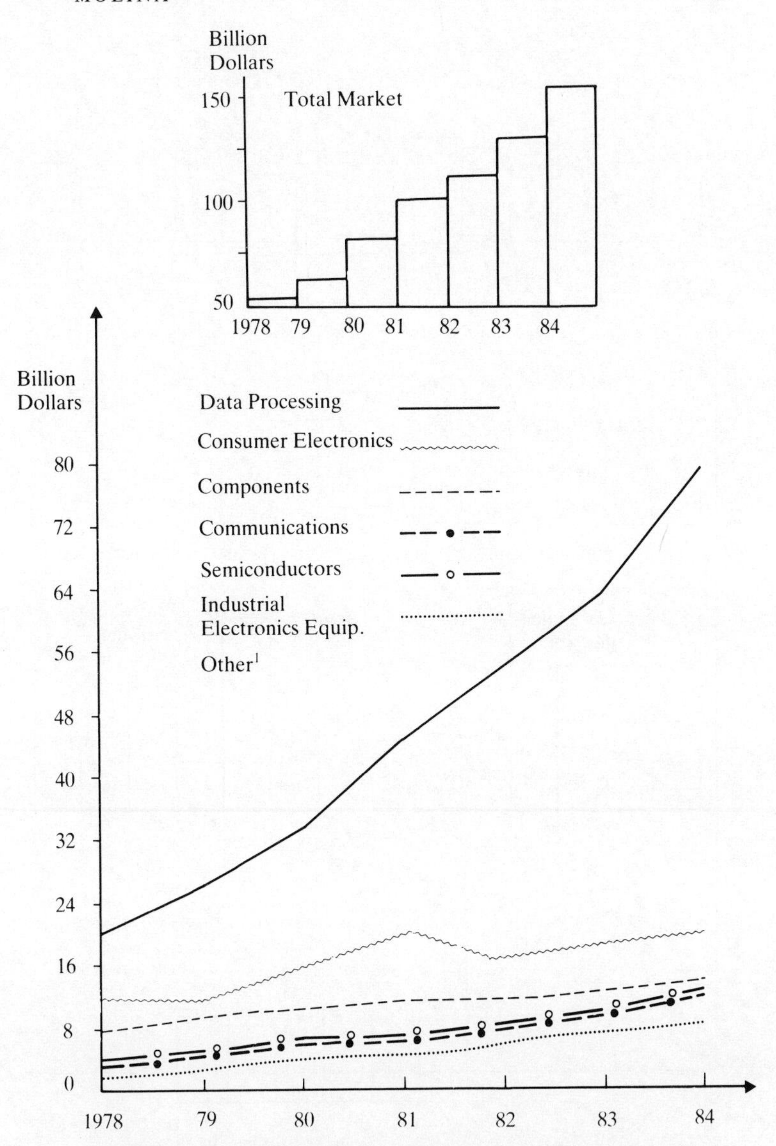

Figure 24. Total US Market for Electronics Equipment and Components (1978-84)

[1] Test and Measuring Equipment, and Medical Equipment not shown in the graph because it overlaps with Semiconductors and Communications

Source. Based on figures given in Molina (1987), p.260

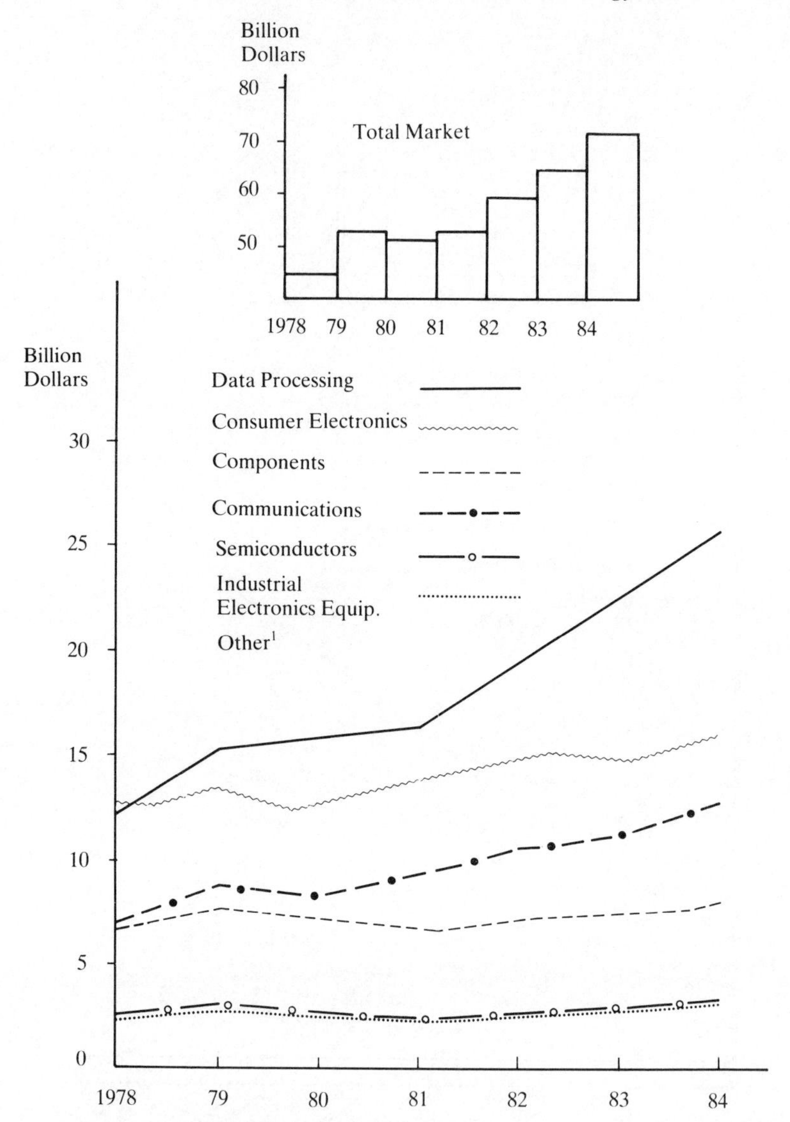

Figure 25. Total European Market for Electronic Equipment and Components (1978-84)

[1] Test and Measuring Equipment, and Medical Equipment not shown in the graph because it overlaps with Semiconductors

Source. Based on figures given in Molina (1987), p.261

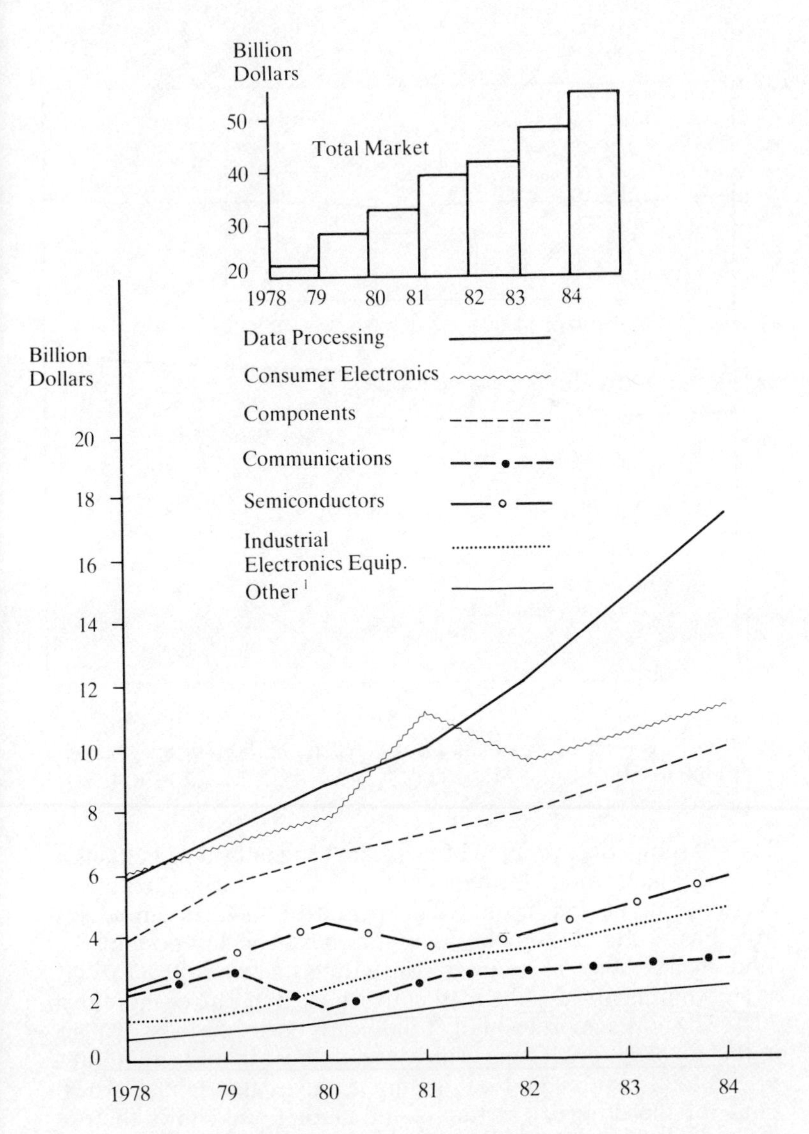

Figure 26. Total Japanese Market for Electronic Equipment and Components (1978-84)

[1] Test and Measuring Equipment, and Medical Equipment

Source. Based on figures given in Molina (1987), p.262.

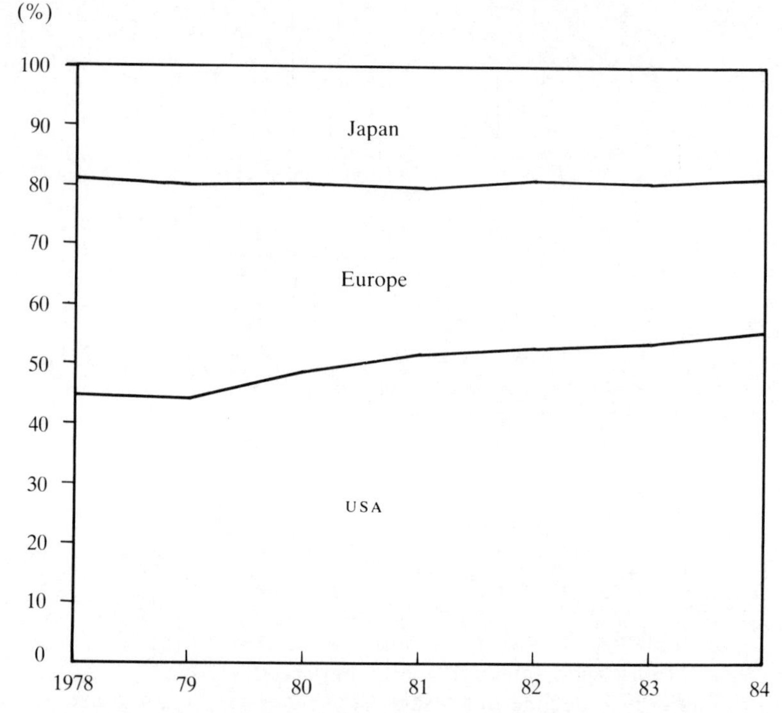

Figure 27. US, European and Japanese Share of their Total Combined Electronics Market (1978-84)
Source. Based on figures given in Molina (1987), p.263

22 to 29 shows the evolution of the overall and individual markets of the three main world electronics powers, namely the United States, Western Europe and Japan, for the period 1978-84. As can be seen from Figure 22, the combined US, European and Japanese market more than doubled in six years. This means that from 1978 to 1984, as shown in Figure 29, the market expanded at a rate of more than 20%. The most important market and clearly one of the most dynamic is that of data processing which accounted for more than 40% of the total market in 1984. Figure 23 gives the annual growth rates for the different sectors of the market and shows that the electronic industry and, particularly, the semiconductor industry spearheading the microrevolution are not free from recessions. The period 1978-9, for instance, was one of very high growth for most sectors; the next year, however, this slowed, and 1980-1 was particularly bad for the semiconductor industry with a negative

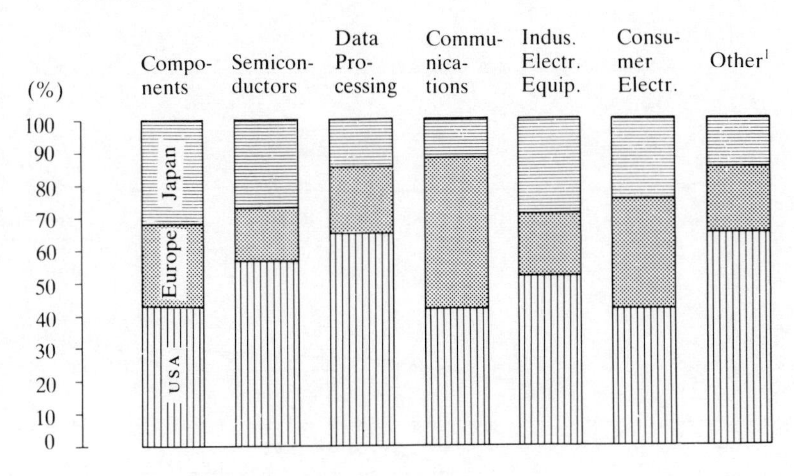

Figure 28. US, European and Japanese Shares of their Total Electronics Market, by Sector (1984)

[1] Test and Measuring Equipment and Medical Equipment

Source. Based on figures given in Molina (1987), p.264

growth of 8.2%. In 1985, this industry was hit by its worst recession ever as worldwide sales fell by an unprecedented 17% with an estimated 29% decline in US sales (*Financial Times*, 6 December 1985 and 18 December 1986).

The overall picture hides differences in size and dynamism between the three main electronics powers. This is provided by Figures 24 to 29 where the respective volumes, shares and growth rates of the United States, Japanese and European markets are shown. Thus Figures 24 to 26 give the growth in volume of each of these markets, overall, and by sector, in millions of dollars. Figure 27 shows the share of the total electronics market accounted for by the US, Europe and Japan respectively. Clearly the most important market is the US's taking more than half of the total combined market for these countries. It is also the most dynamic, having increased its share by approximately 10% between 1978 and 1984. Europe is the second largest market with a share of 25.6% of the total combined market in 1984, but it is clearly the least dynamic, having dropped more than ten points from a share of 36.6% in 1978. The Japanese market with a share of 19.5% of the total combined market has increased by about 1% only since 1978. It is still third, but in 1984 the gap with Europe had been closed to only 6%.

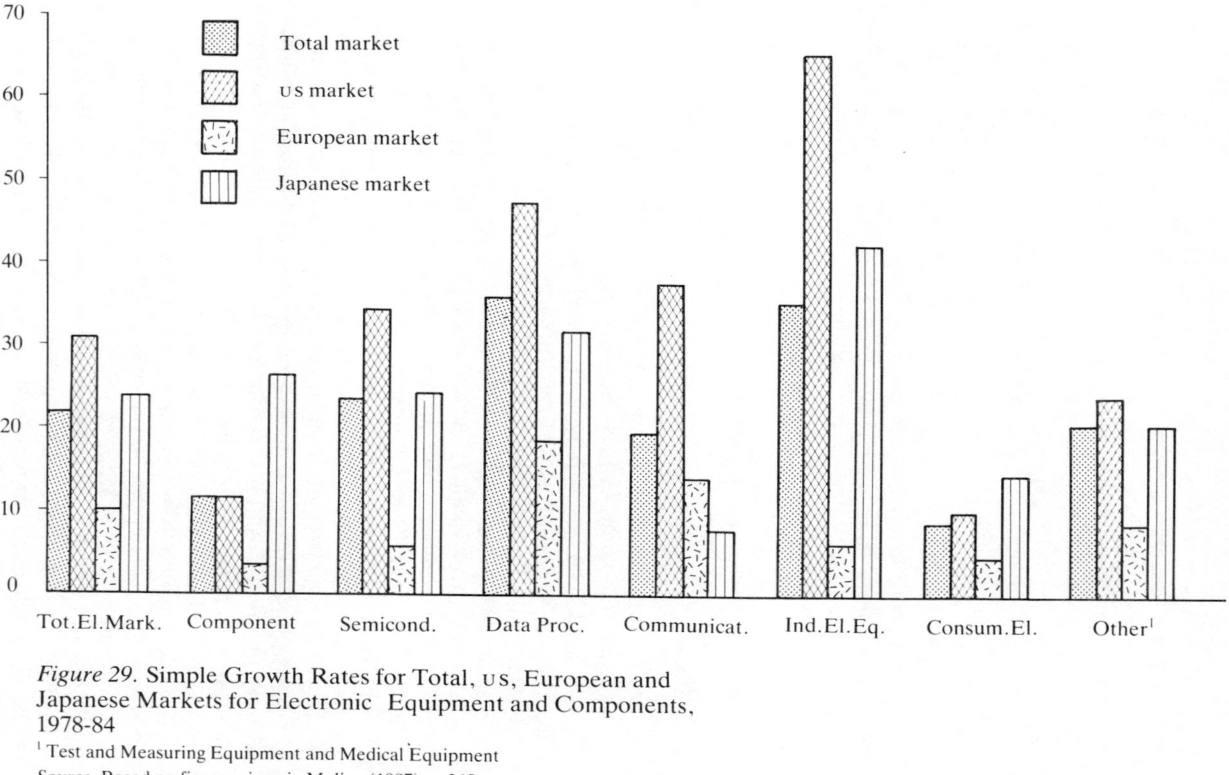

Figure 29. Simple Growth Rates for Total, US, European and Japanese Markets for Electronic Equipment and Components, 1978-84

[1] Test and Measuring Equipment and Medical Equipment

Source. Based on figures given in Molina (1987), p.265

The comparative position of the US, European and Japanese markets by sector in 1984 is given in Figure 28 where the US market's dominance can be seen, not only in data processing but also for semiconductors and the 'other' category, containing test and measuring equipment and medical equipment. The European market is particularly strong in communications where it accounts for 46% of the total, while the Japanese market is particularly strong in components, semiconductors and industrial electronics equipment. Individually, the internal structure of these major markets also show significant differences. In the United States, for instance, the data processing market has a much greater relative weight in the total US market than it does in Europe and Japan. On the other hand, in Japan the consumer electronics market has until recently played a greater relative role in the country's total market than has been the case in the US and Europe, while in Europe, the relative weight of the telecommunications market is far greater than that shown by the same market in the US and Japan. This is interesting since, corroborating the importance of the size of markets for technological development, we shall later see that these individual market features coincide with each region's particular strength. For instance, we shall see that it is precisely in telecommunications where the European electronics industry possesses greatest strength in comparison with Japan and the US. For Japan, it is mainly in semiconductors and industrial electronic equipment that her challenge to the US electronics industry has been most impressive, with strong efforts being directed towards the computer market where the US supremacy is strongest.[15] Figure 29 corroborates this by showing the dynamism of the respective US, European and Japanese markets by sector for the period 1978-84. The simple growth rate for the Japanese market in semiconductors is highest, while its growth in industrial electronics was almost equal to that of the US market. The Japanese market also grew faster in components and test and measuring equipment, but the US market had no parallel in data processing and telecommunications. Apart from telecommunications where Japanese growth was the smallest, the European market's growth rate lagged far behind in every sector. The semiconductors sector was particularly poor with only 5.8%. Figure 25 shows that Europe's market, after falling in 1979, had barely recovered by 1984.

In all, the US is by far the dominant market with Europe in second place but losing ground. In contrast, Japan's market is far more dynamic, growing even faster than the US's in some areas. In the crucial sectors of data processing and telecommunications,

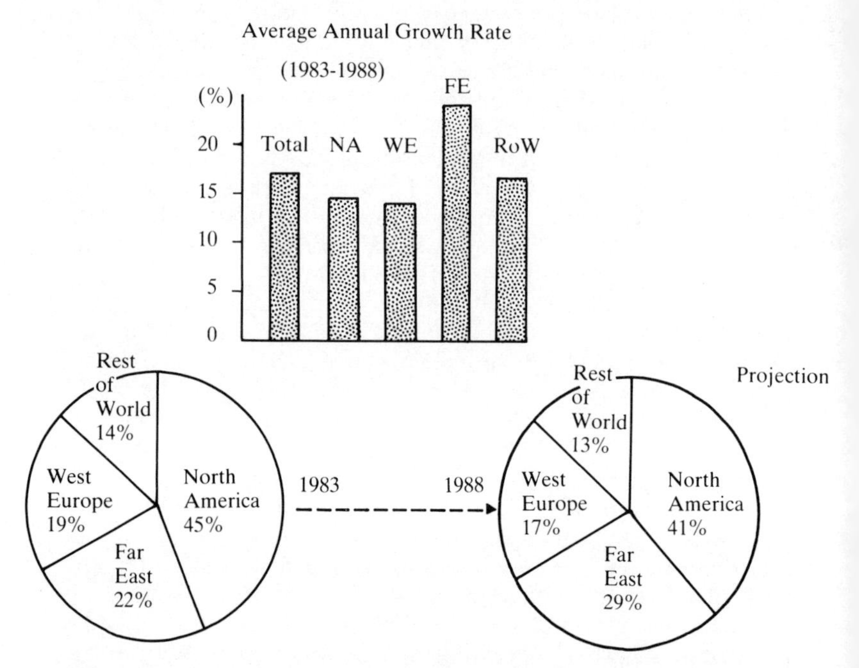

Figure 30. Share of Worldwide Production of Electronic Equipment and Average Annual Growth Rate by Main Regions, 1983 and 1988
Source. Based on figures given in *Electronics Week*, 1 January 1985, p.62

the US market is clearly the most dynamic while the Japanese has grown faster in semiconductors and at a similar high rate as the United States in industrial electronics.

b) Current Trends in the Global Battle for Control of Electronics Markets

Statistical analysis of the growth of the electronics market does not answer the crucial question of who the dominant and competing forces are, nor how tightly the electronics market is concentrated and controlled. Figure 30 sheds some light, showing the share of (worldwide) production of electronic equipment and average annual growth rate for the period 1983-8 by main regions of the world. According to this, in 1983, North America was leader with 45% of world production, followed by the Far East with 22% and Europe in third place with 19%. In terms of electronics production, however, Figure 30 clearly reveals the Far East as the most dynamic region with an average annual rate of nearly 24% for the period 1983-8. Least dynamic is Western Europe with a rate of less than 14%, while North America manages to top the 14% mark. The Far East's share of worldwide production of electronics equipment is expected to increase to 29% by 1988, whereas North America's share is expected to fall to 41%, and Western Europe's share to fall from 19% to 17%. During the same period, 1983-8, electronics production for the rest of the world is expected to change only slightly from 14% to 13%. In a more detailed picture, Figures 31 to 35 and Table 10 show the position of the world's leading companies for the critical sectors of semiconductors and semiconductor equipment, computers, telecommunications, CAD/CAM and robots.

Although the US is dominant in the electronics industry overall, this does not mean that such dominance is uniformly reflected throughout the different sectors of the industry. In fact, Figures 31 to 35 and Table 10 reveal a varying pattern, with Japan strongly challenging US dominance in semiconductors and robotics. Moreover, Figures 36 and 37 indicate that the challenge is also strong in the field of NC machine tools and CNC lathes.[16] In the field of semiconductors,[17] for instance, in 1984 two United States firms were among the industry's three world leaders, but three Japanese companies were among the first five. Most importantly, Japanese firms were rapidly gaining ground against US firms.[18] Thus in 1980, National Semiconductor (US) was third in the league while the production of NEC the main Japanese semiconductor firm, was only about half that of TI the leading US firm (NEC $769 million against TI $1,580 million) (Bessant, 1984). In 1985, however, for the first time a Japanese company, NEC, became the leader of the world market (Molina, 1986), while a year later, as illustrated in Figure 31, three Japanese companies — NEC, Hitachi and Toshiba — had captured the top of the semiconductor league. The American company TI, long-standing world leader, had been relegated to fourth place with Motorola fifth. In all, US companies still had 45% of the world market for semiconductors in 1985 with Japan accounting for 42% (*The Economist*, 7 March 1987). However, some observers estimate that by the late 1980s Japanese companies will have captured 50% of the market (*Electronics*, 9 February 1984).[19]

This comparison is tempered by the fact that while Japanese figures include all the country's chip production, US figures exclude production for internal consumption by both IBM and AT&T. It is estimated that both these companies produce about $2.5 billion of chips a year for their own use.

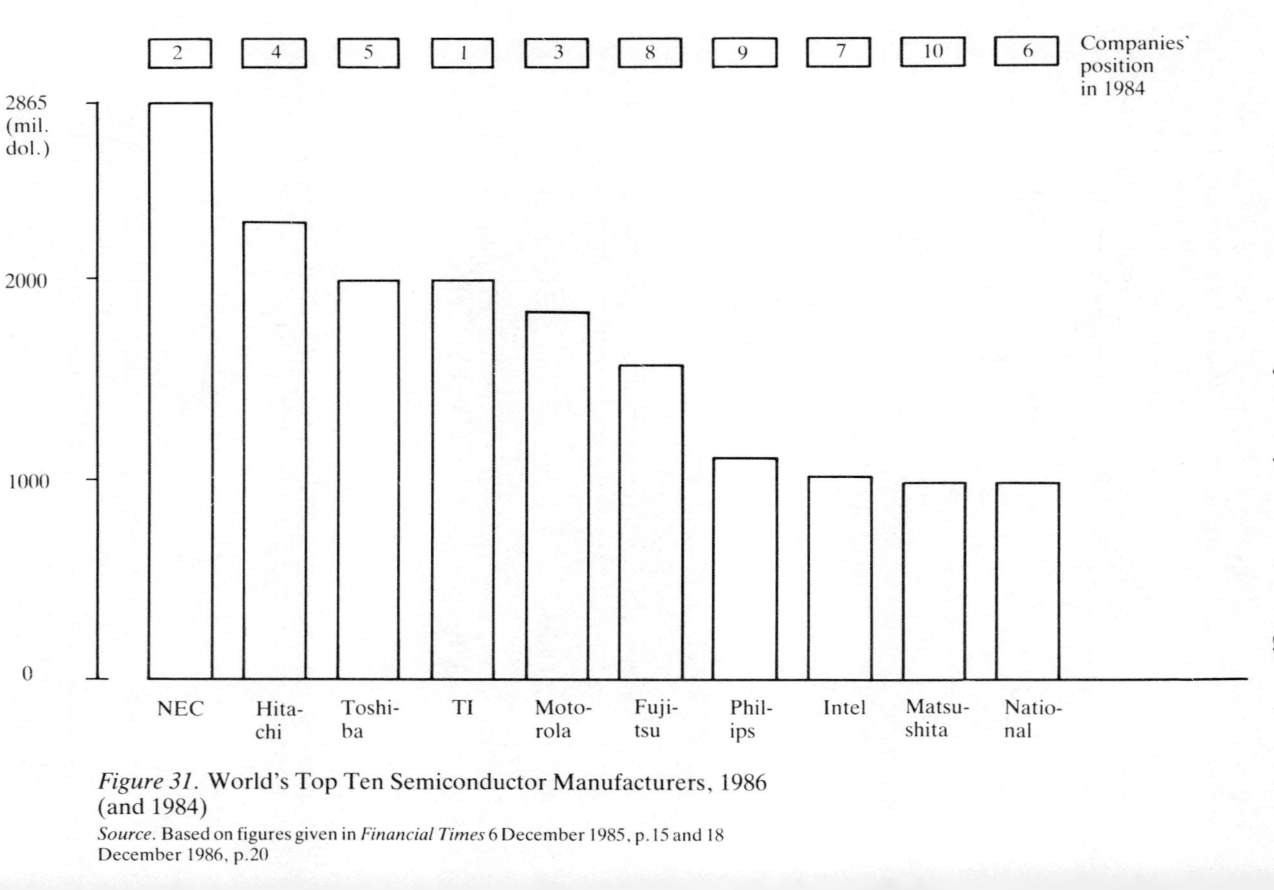

Figure 31. World's Top Ten Semiconductor Manufacturers, 1986 (and 1984)

Source. Based on figures given in *Financial Times* 6 December 1985, p.15 and 18 December 1986, p.20

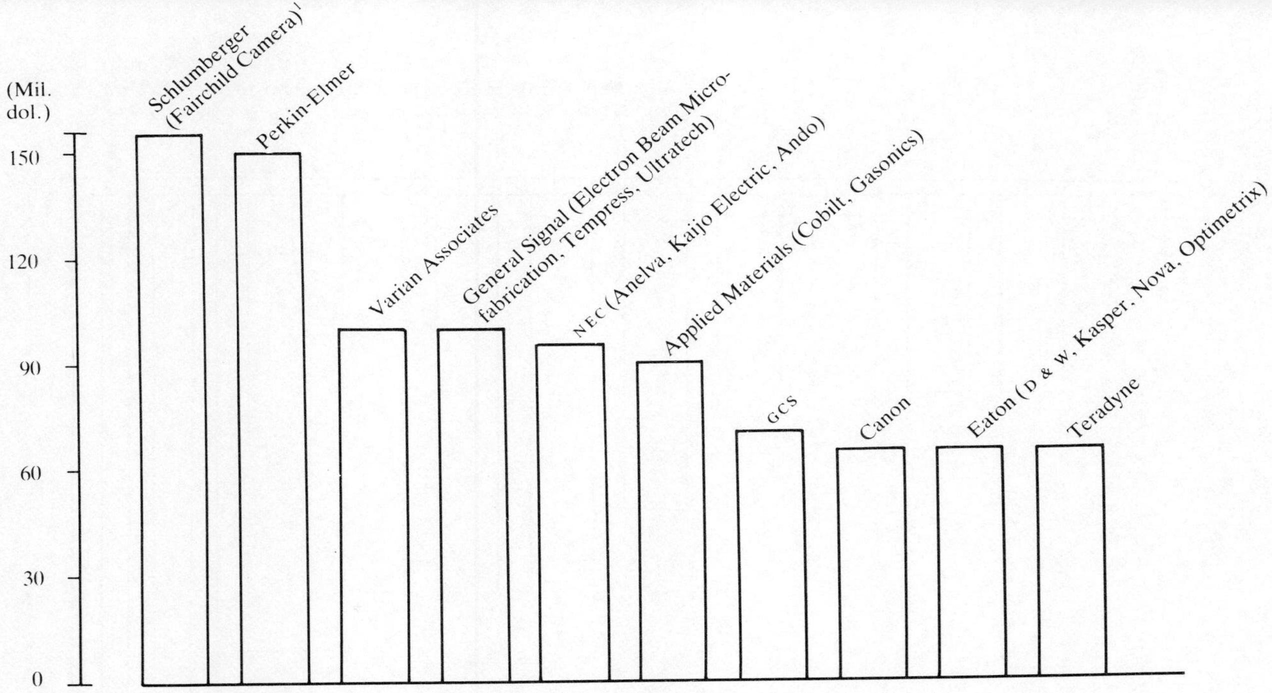

(Mil. dol.)
150
120
90
60
30
0
Schlumberger (Fairchild Camera)[1]
Perkin-Elmer
Varian Associates
General Signal (Electron Beam Micro-fabrication, Tempress, Ultratech)
NEC (Anelva, Kaijo Electric, Ando)
Applied Materials (Cobilt, Gasonics)
GCS
Canon
Eaton (D & W, Kasper, Nova, Optimetrix)
Teradyne

Figure 32. World's Top Ten Semiconductor Equipment Companies and Sales: 1982

Source. Based on figures given in K. Rothschild (1983), p.4

[1] Fairchild Camera has now been acquired by National Semiconductor

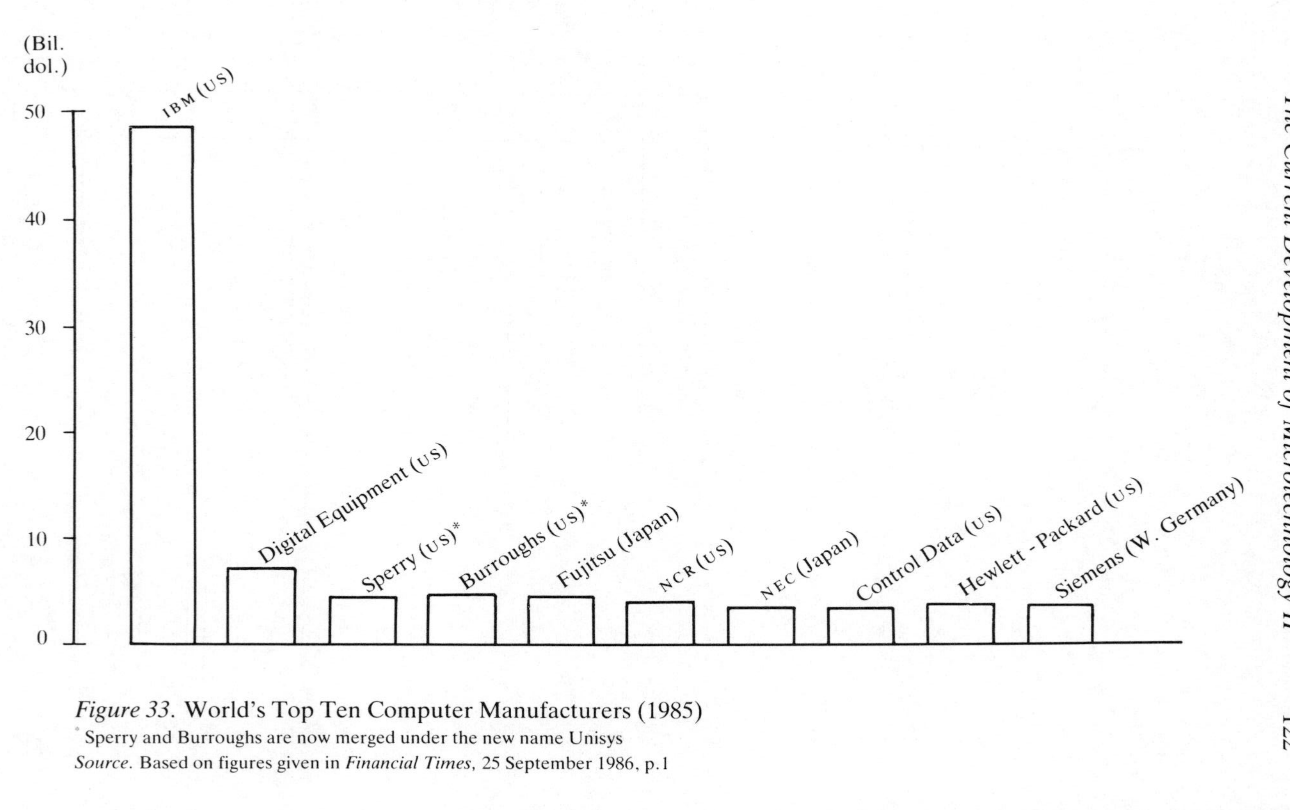

Figure 33. World's Top Ten Computer Manufacturers (1985)

* Sperry and Burroughs are now merged under the new name Unisys

Source. Based on figures given in *Financial Times*, 25 September 1986, p.1

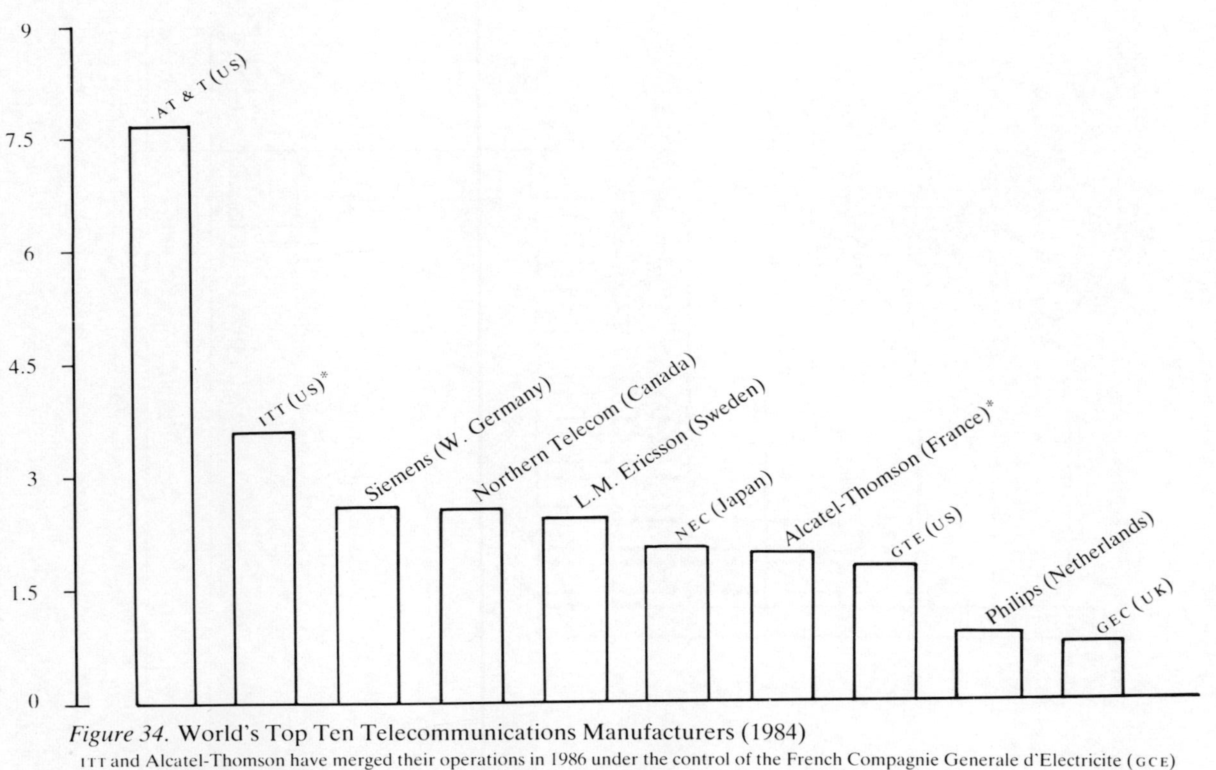

Figure 34. World's Top Ten Telecommunications Manufacturers (1984)

ITT and Alcatel-Thomson have merged their operations in 1986 under the control of the French Compagnie Generale d'Electricite (GCE)

Source. Based on figures given in *Financial Times*, 1 December 1986, p.I (Survey)

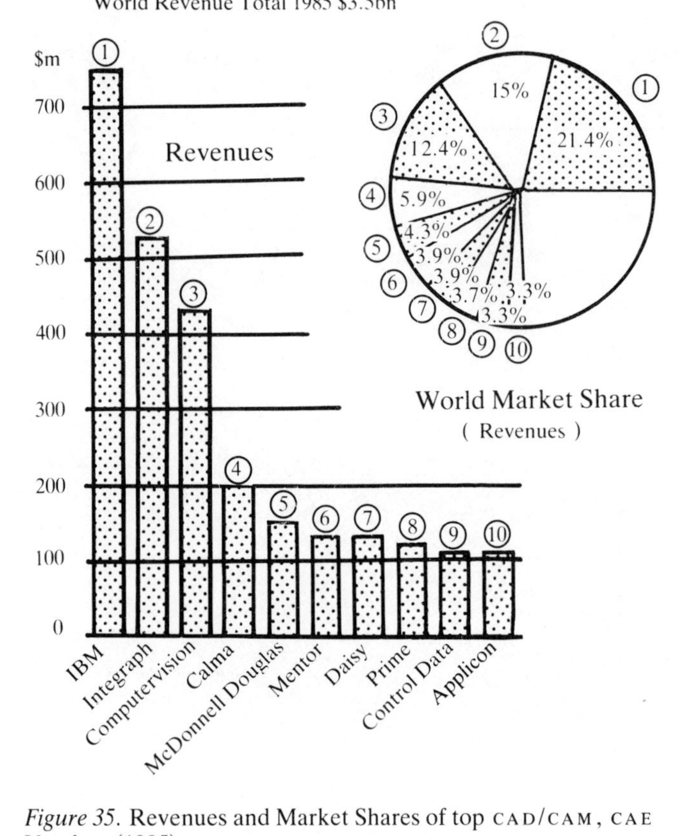

Figure 35. Revenues and Market Shares of top CAD/CAM, CAE Vendors (1985)

Source. Financial Times, 3 February 1985, p. VIII (Survey)

Table 10: Market Shares of Major Robot Manufacturers (1981)

Firm	USA% ($250m)	Europe% ($230m)	World%
Unimation	32	24	16
Cincinnati	32	7.4	3
ASEA	—	24	13
Kawasaki	—	—	14
Fanuc	—	—	13

Mitsubishi	—	—	2
Yasukawa	—	—	7
Hitachi	—	—	12
Kuka (FRG)		19.1	
Devil Biss (Norway)	5	14.1	
Other	5	11.4	18

(Sources: CSI)

Source: J. Bessant (1984), p.35

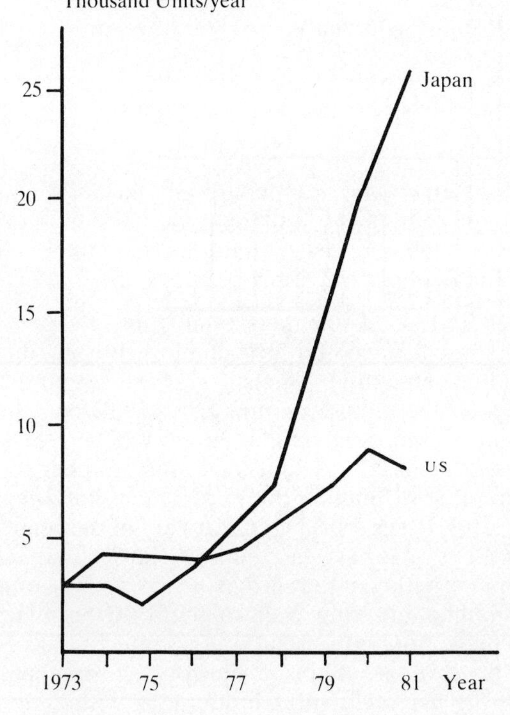

Figure 36. Evolution of NC Machine Tool Production in Japan (1973-81)

Source. J. Abegglen and A. Etori (1983), p.J18

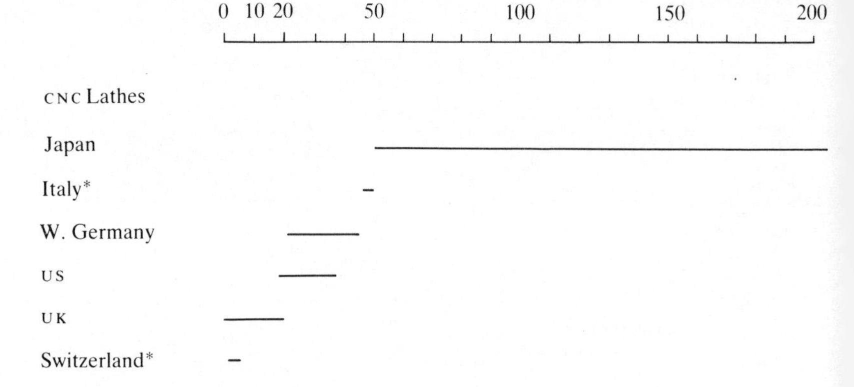

Figure 37. Range of Monthly Production Volumes for CNC Lathes (1982-3)

Source. E. Sciberras and B. Payne (1984), p.120

* Both the Italian and Swiss samples included only one CNC lathe manufacturer. So a 'range' could not be indicated.

> If all these captive American chips are included, Japan probably has only about a 30% share of the world market put at more than $30 billion a year. America's share may well be over 50% (*The Economist,* 7 March 1987, p.73).

For the beleaguered US chipmakers competing in the open market, however, these figures offer little comfort. Indeed, during the recent recession, even more pessimistic views were expressed, suggesting that if Japan continues to gain ground 'the companies that created the world semiconductor industry will be relegated to the role of chip designers — no longer playing a principal part in the manufacture of semiconductor devices' (*Financial Times,* 5 December 1985, p.16).[20] The worldwide recovery of the semiconductor market since 1987 may have eased this pessimism for the time being, but US chipmakers know that if they want to compete with the Japanese in the long term they need to pool their resources in critical areas and make programmes such as Sematech a success.

In contrast, in the strategic sector of semiconductor equipment US domination of the market is still considerable, with eight out of the top ten companies in 1982 being US based (Figure 32). The leader Fairchild, however, was owned by the French company Schlumberger between 1979 and 1987. The new owner is National Semiconductor who paid $122 million after a $240 million bid by

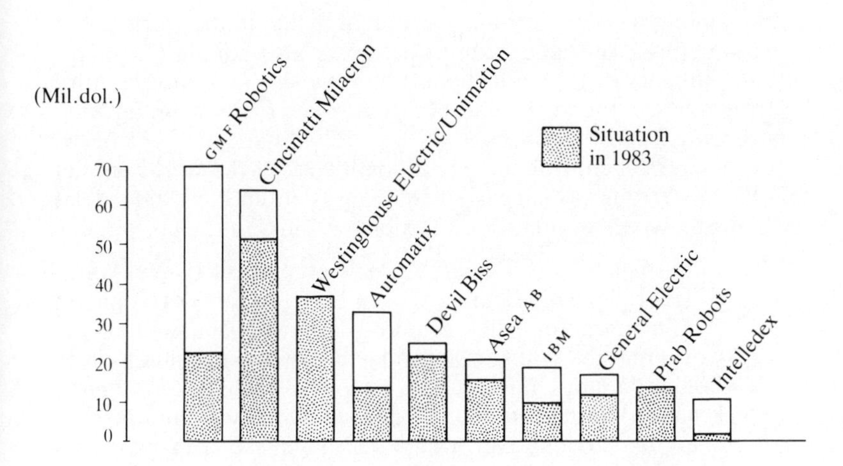

Figure 38. Estimated Sales by Top Ten US Robot Vendors: 1984 (and 1983)

Source. Based on figures given in *Electronics Week* 23 July 1984, p.96

the Japanese company Fujitsu was blocked by the US government on grounds of national security. Two of the ten companies in Figure 32 were Japanese and their continuous advance was causing some anxiety to US manufacturers.[21] Thus, in 1984 the United States' share of the market for advanced equipment in Japan had dropped to 32% and eleven Japanese companies were 'establishing beachheads in the US expecting to sell $125 million of machinery' (Uttal, 1984).[22]

In the field of robotics, Japan's position is powerful, five of its companies being within the world's top ten and holding almost 50% of the world market in 1981. In this sector the US had two leading companies in 1981 but, by 1983, some large corporations were entering the market, among them IBM, United Technologies, Westinghouse, GM and GE. Most entered the robot market through acquisitions and licensing agreements with established producers. Thus, Westinghouse acquired Unimation, GM made a licensing agreement with Fanuc (Japan) and GE obtained licences from both Renault (France) and Volkswagen (Germany) (*The Economist*, 4 June 1983). Figure 38 shows that these moves have already affected the picture of the US robot market. In 1984, market leadership had fallen to GMF Robotics Corp., the GM-Fanuc joint company, while Unimation, now Westinghouse Electric/Unimation Inc. had been displaced to third place. Also, IBM and GE appeared for the first time among the top ten robot vendors. Finally, in the world robot market, three European companies were also among the major robot builders in 1981. Of them, the Swedish ASEA had the third largest share of the world market together with Fanuc from Japan,[23] both with 10% of the market.

In numerical control, Japanese dominance of the world market is equally strong, as suggested by Figures 36 and 37. Indeed, as Jacobsson writes in relation to flexible NC units,

> In particular, one Japanese producer, Fujitsu Fanuc, which collaborates with Siemens . . . in both product development and in marketing, has acquired around 50 percent of the world market. Thus total production of NC machine tools in 1980 in Japan, United States, Federal Republic of Germany, France, United Kingdom, Italy and Sweden amounted to over 40,000 units, of which the Japanese firm produced 21,000 NC units. Other producers of significance are Siemens, General Electric and Olivetti as well as a number of machine tool producers which supply their own NC units (Jacobsson, 1982, p.50).[24]

Fanuc's dominance was still strong in 1986. It is estimated that the company supplies 45% to 50% of all CNC systems manufactured in the world and about 65% to 70% of all those manufactured in Japan (*Financial Times*, February 1986). With process control equipment, however, the situation changes in favour of US and European firms. In fact, the Japanese world market share in this category was only 12.7% in 1980. US and European shares were 43.3% and 29.3% respectively. In addition, twelve US and European companies with annual sales ranging from $200 to $750 million controlled 62% of the world market while only three Japanese firms, all with technical links with firms in OECD countries, were in competition with sales of about $200 to $300 million each (Jacobsson, 1982).

In the software-intensive CAD/CAM sector, the picture seems to be one of US dominance of the world market, although figures are not available for Japanese and European sales of these systems (Kaplinsky, 1982). Nevertheless, the US market was dominant in 1985 with Europe far behind. Indeed, it has been estimated that the entire CAD/CAM installed base in Germany is the same as that of Boeing (*Financial Times*, 16 July 1985). By the late 1970s, however, European and Japanese markets were growing faster than the US market, and the Japanese had become the fastest growing single market for CAD systems sold by the US turnkey suppliers (Kaplinsky, 1982). By this time, no Japanese company, apparently, had entered the market but there appeared to be production within the country for internal consumption by large companies such as Matsushita (*ibid.*). In 1982, however, *Datamation* reported that the Japanese had begun moving into CAD the previous year, and names NSC, Mitsubishi and Fujitsu as three of the Japanese companies developing and/or offering some CAD systems (*Datamation*, December 1982). In addition, one Japanese firm, Yokogawa Electric Works, had obtained an exclusive licence from Gerber (US) to manufacture and market CAD systems in Japan, Korea and Singapore (Kaplinsky, 1982). Figure 35 shows the US market shares for the leading companies in the US CAD/CAM market and, probably, in the world market too. As we can see, IBM has taken the market lead,[25] and it is estimated that, by 1985, its share will increase to about 25% of the total market.

In telecommunications, the US and European position in the world market is overwhelming (Figure 34). Japan had only one firm in the top ten league in 1984 and its share was only 7.8% of the combined sales of the ten top companies. US companies accounted for almost 50% of the market and by far the biggest company, AT&T Technologies, alone claimed almost 30% of the market. In turn, three European firms were high in the league with 33% of the world market. Europe's strength in this field is commensurate with the huge importance of its telecommunications market and the fact that in most countries this has been until recently a market 'controlled by powerful state monopolies that buy mainly from favoured national manufacturers' (*Newsweek*, 23 May 1983, p.14). An interesting change in this market is the recent acquisition of ITT's telecommunications operations by the French Compagnie Générale d'Electricité, the parent company of Alcatel-Thomson (*Financial Times*, 4 and 7 July 1986). This puts France second to AT&T with approximately 21% of the telecommunications market. The US share is now down to about 36% whereas the combined European force has become dominant with a 47% share of the ten top companies' telecommunications sales.

In the computer sector, the United States heavily dominate the world market. As Figure 33 shows, in 1985, seven of the ten largest computer manufacturers were US companies. In addition, the first four companies were all from the US[26] with the leader IBM holding alone around 60% of the world market (Gregory, 1983). The combined share of the world market for all US companies was approximately 80% in 1984, with the Japanese share less than 10% (*Business Week*, 16 July 1984). In this sector, however, the Japanese have been making strenuous efforts to make greater inroads in the world market and, eventually, to break IBM's stranglehold.[27] For a time, it seemed that the Japanese were succeeding, with Fujitsu, Hitachi and NEC beginning to produce mainframe computers at least as powerful as IBM's top machines. In 1979, Fujitsu overtook IBM as the leading computer company in the Japanese market, leading some commentators to speculate that by 1990 Japan would account for 30% of the world market (Gregory, 1983).[28] This possibility has not materialised however, and, if anything, there has been an aggressive response by IBM which, according to some observers, has forced the Japanese on the defensive in their home market. The problem seems to lie in the software base accumulated by Japanese companies which is weak in comparison with the US base.[29] This has meant that both Fujitsu and Hitachi began making their computers compatible with IBM's and, consequently, have come to depend heavily upon IBM's operating-system software.[30] In 1982, however, IBM decided to take advantage of its software muscle to counteract the Japanese drive. It filed a suit against Hitachi for allegedly stealing IBM software and decided to stop disclosing the source code describing the operating-software on which all applications programmes must run. As a result, both Hitachi and Fujitsu began to divert huge amounts of resources into software to avoid infringing IBM's copyrights. Without the source code, some analysts argued that Hitachi and Fujitsu would find it impossible to produce operating-system software that can compete with IBM's new operating systems (*Business Week*, 16 July 1984).

The first reaction from Japan was to propose a change in the country's copyright law that would enable MITI to force a company to license software to its Japanese competitors. Opposition from the US, however, halted this move, but, after five years, the dispute has been settled with an agreement that permits Fujitsu to examine proprietary IBM mainframe systems software in exchange for cash payments (*IEEE Spectrum*, November 1987). Nevertheless, IBM's move against the Japanese has pushed them into a major software effort, having shown them the dangers of heavy dependence upon competitors.[31] In this context the software effort of the Fifth Generation Computer project acquires even greater importance, as does the Japanese collaboration with AT&T to develop an alternative software to IBM's on the basis of AT&T's UNIX operating system. According to MITI, the plan would entail a five-year effort with AT&T, to foster software development within Japan. It is expected to cost some $125 million and requires AT&T to work with several Japanese computer companies on the development of the UNIX operating system (*Datamation,* 1 October 1984).[32] Thus, IBM may have triggered an alliance of interests between its most formidable opponents.

In summary, the US is the major force in world electronics. However, such dominance is not total, Japan is leading in such sectors as memory semiconductors, robots and numerical control. Thus far, software has proven to be Japan's Achilles' heel, preventing it from making significant inroads in software-intensive markets such as CAD/CAM and computers — the most important of all. In the computer markets, US dominance can be considered complete; Japan is likely to challenge this only in the 1990s if its software effort, now in progress, is productive. Apart from its strong presence in the communications sector Europe is lagging increasingly far behind in this international competition, unable to sustain the demands and the pace of advance partly imposed by the US-Japan competition. Also evident from the above analysis is the concentration and control of all main markets by a handful of companies. In general, the control exercised by the ten or so largest companies is overwhelming and, in the cases of computers and numerical control one company alone controlled 60% and 50% of the world market respectively. Within this context, the field of semiconductors remained less concentrated, with the top eight companies controlling about 50% and the top twenty about 75% of the world market (Levin, 1982). Even here, however, general trends are pushing for greater concentration as the convergence of technologies stimulates market integration, thus raising the barriers to entry into the industry. As one major study concluded,

> Even those already in the industry are finding it extremely difficult to pursue the constantly advancing technology because of rapidly rising R&D and equipment costs. This condition, if continued, could lead to an industry dominated by 5 to 10 large vertically integrated producers (US Department of Commerce, 1979. p.103).

The same trend is seen throughout the electronics industry. In telecommunications, for instance, 'it is possible that by 1990 only 4 main manufacturers will be left in the field (1 American, 1 Japanese and 2 Europeans)' (Rada, 1980, p.115). For the electronics industry as a whole, Bessant (1984) predicts that 'a few major firms, highly integrated in both vertical and horizontal ways are coming to dominate the IT industry — perhaps as few as 40 firms worldwide' (Bessant, p. 57).

c) Global Alliances in the Battle for Control of the Electronics Infrastructure.

As technology convergence continues, and the struggle for the control of the electronics infrastructure intensifies, there is every reason to believe that the concentration of capital in the electronics industry will reach the proportions predicted above. The most dramatic illustration of this trend has been the wave of company acquisitions, joint ventures and other forms of company interlinking within the world electronics industry. Even the largest companies have been unable rapidly to accumulate the resources necessary to succeed in the worldwide battle for control of the electronics markets.[33] In the critical economic conditions of recent years, mergers and takeover activity have increased markedly within the economy as a whole, but, as Figure 39 illustrates, within the information technology sector the pace has been faster. In addition, within the electronics industry, companies are interlinking, not just through mergers and takeovers, but in a variety of other ways. In electronics, therefore, the cause behind capital concentration is more profound than elsewhere. Above all, it suggests that the development and production of entire systems (e.g. office automation) will play a decisive role in the control of the industry.

One reason is simply that, at R&D and production levels, advances in one area of electronics technology tend to feed all others. Another reason is that all the largest companies in the electronics field are already integrated to an important degree. The success of Japanese companies such as NEC, Hitachi, Fujitsu, has been significantly associated with their integration.[34] This has been one of the keys to Japanese advance in semiconductors, since these companies not only have an internal demand for the components they produce, but also have semiconductors as just one part of their business. As a result they have been more sheltered against the market fluctuations which have affected the US semiconductor industry.[35] Table 11 shows the characteristics of the six dominant firms in the Japanese semiconductor industry. As can be seen, the percentage of semiconductors sales to total sales is quite low for them all. They are all integrated into other electronic products and, if we look at the figures showing the world leaders in the different electronics sectors, we find that the same names tend to

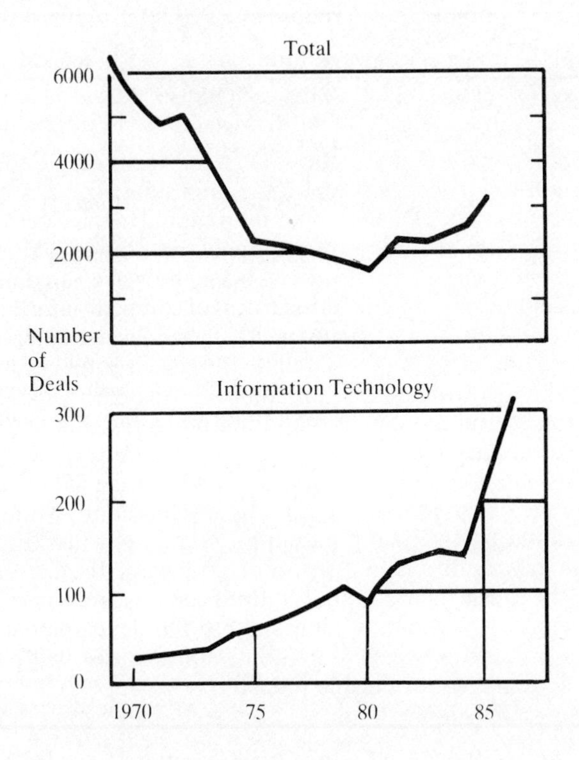

Figure 39. Pattern of US Merger and Acquisition Activity, 1970-86
Source. Financial Times, 5 March 1987 p.8

recur. NEC, for instance, is found in semiconductors, computers and telecommunications, while Fujitsu is found in semiconductors, computers and, through its shareholding in Fanuc, in robots and numerical control. Such companies, according to Ferguson (1983), 'are in a far better position to take advantage of advances in microelectronics' (Ferguson, p.28).

In the US, both IBM and AT&T have always been vertically integrated companies and are rapidly following the Japanese path by integrating horizontally into almost every main sector of the electronics market. Indeed, IBM, apart from its leadership in computers and CAD/CAM, is also active in telecommunications and

Table 11: Characteristics of Dominant Firms in Japanese Semiconductor (SC) Industry

Firm	Total Sales ($bil.)	SC Sales ($mil.)	SC Sales % Total Sales	SC Strength	Systems Market
NEC	3.3	590	17.8	MOS-LSI (NMOS, CMOS) Memory (16K strong, 64K redesign strong) MPUs (4 bit)	Leading IC Power-house; Leader in Telecommunications Technology, Linear Computers
Hitachi	10.7	440	4.1	MOS-LSI (NMOS, CMOS) Memory (16K, 64K very strong?); Bipolar logic – ECL; Shottky – TTL; MPU-Motorola	Leading Diversified Systems, Computers, producer in communications, consumer, heavy industrial and electrical machinery
Fujitsu	1.8	120	6.7	MOS Memory (NMOS) (64K Strong); Bipolar Logic-ECL	Leader in Computers
Toshiba	7.1	390	5.5	CMOS, MSI-LSI (16K static) (64K) Consumer linear CMOS-MPU	Diversified Systems, esp. consumer, bus-systems, instrumentation, appliances and electrical equipment
Mitsubishi	3.9	150	3.8	Industrial/Consumer linear; 64K RAM entrant; some ECL	Diversified Systems, small bus, computers, industrial & heavy electrical equipment, medium appliances.
Matsushita	9.8	125	2.3	Consumer linear New 64K static RAM (strategy shift)	Leader in consumer & appliances, home computers

Source: M. Borrus *et al.* (1982), pp.66-7

robots. This trend has forced other companies to follow, seeing a wave of acquisitions and collaboration agreements[36] as they try to improve their resource-base in order to cope with the increasing range of demands imposed by worldwide competition.[37] As has been stated,

Forming alliances . . .is becoming a critical element in almost any successful strategy . . . Without this kind of cooperation, many manufacturers will not survive the restructuring (*Business Week*, 16 July 1984, p.49).

Tables 12 to 15 indicate the type of arrangements and the extent of capital concentration taking place. The scale of this process is immediately obvious. For example, Table 12 lists

Table 12: Corporate Investment in US Semiconductor Companies

US Semiconductor Company	Corporate Investor	Percent Ownership[1]	National Base
Advanced Micro Devices	Siemens	20 %	West Germany
American Microsystems	Robert Bosch	12.5%	West Germany
	Borg Wagner	12.5%	US
Analog Devices	Standard Oil of Indiana		US
Electronic Arrays	Nippon Electric		Japan
Exar	Toyo	53 %	Japan
Fairchild Camera	Schlumberger		Netherlands Antilles
Frontier	Commodore International		Bahamas
Inmos	National Enterprise Board		United Kingdom
Interdesign	Ferranti		United Kingdom
Intersil	Northern Telecom	24 %	Canada
Litronix	Siemens		West Germany
Maruman IC	Toshiba		Japan
Micropower Systems	Seiko		Japan
Monolithic Memories	Northern Telecom		Canada
MOS Technology	Commodore International	12.4 %	Bahamas
Mostek	United Technologies[2]		US
Precision Monolithics	Bourns		US
Semi, Inc.	General Tel. & Elec.		US
Semtech	Signal Companies		US
Signetics	Philips	Merger	Netherlands
Siliconix	Electronic Engr. of Calif.		US
	Lucas Industries	24 %	United Kingdom
Solid State Scientific	VDO Adolf Schindling	25 %	West Germany
Spectronics	Honeywell		US
Supertex	Investment Group		Hong Kong
Synertek	Honeywell		US
Unitrode	Signal Companies		US
Western Digital	Emerson Electric		US
Zilog	Exxon		US

1 No percentage indicates 100% (wholly owned), or presumed to be wholly owned, in the absence of data.
2 Mostek has now been acquired by Thomson of France.

Source: M. Borrus *et al.* (1982), pp.40-1

twenty-eight US semiconductor companies which, by 1982, had been either totally or partially acquired by major corporations.[38] Later, in a highly publicised development, IBM acquired 20% of Intel's shares, the world's leading microprocessor manufacturer. In addition, as Table 13 shows, AT&T alone has made over

Table 13: AT & T's Strategic Alliances Aimed at Broadening its Presence in the Computer Market

Company	Arrangement
Olivetti Corp.	Makes the AT & T 6300, compatible with IBM Corp. Personal Computers
Convergent Technologies Inc.	Makes the AT & T 7300, a personal computer that operates with AT&T's Unix software
Electronic Data Systems Corp.	Joint marketing arrangement for custom-designed computer networking solutions
Quotron Systems Inc.	Joint development and marketing of a computer-based financial-information system for stockbrokers and security analysts
Microsoft Corp.	Joint development of compatible versions of Unix System V and Microsoft's Xenix, which is derived from Unix
Motorola Inc., Intel Corp., Zilog Inc., National Semiconductor Corp.	Development of Unix System V for use with microprocessors made by these vendors
Amdahl Corp.	Agreement to ensure compatibility of Unix System V with Amdahl's UTS software
Omnicad	AT & T acquired an equity position in this company, which sells software for computer-aided design and computer-aided manufacturing; joint marketing of CAD software on AT & T personal computers
Value-added resellers, independent software vendors, original-equipment manufacturers	AT & T has signed over 80 of these firms to help it penetrate niche markets

Source: Electronics, 5 August 1985, p.28

Table 14: US & Japan Semiconductor Agreements (1984-5)

Companies	Date	Technology Exchanged
RCA Corp.-Sharp Corp.	April 1985	Share CMOS design and process technology
Intel Corp.-Toshiba Corp.	Jan. 1985	Toshiba is manufacturing and marketing Intel interference controller chips
Motorola Inc.-Hitachi Ltd	Jan. 1985	Motorola will second-source Hitachi's 6301 microcomputer
Intel-Fujitsu Ltd.	Nov. 1984	Fujitsu will manufacture Intel's 80186 and 80286 microprocessors and the 8051 controller chip
Intel-Fujitsu	Sept. 1984	Agreement includes Intel logic design and Fujitsu layout for a new 8/16-bit I/O processor known as the 8089-2
Motorola-Toko Electric Co.	July 1984	Subcontracted bipolar fabrication in Japan
Standard Microsystems Inc.-Oki Electric Industrial Co.	July 1984	Non-exclusive cross-licensing of each company's patents and patent applications
Monolithic Memories Inc.-Fujitsu	July 1984	Monolithic Memories will second-source Fujitsu's B-series bipolar TTL gate arrays
Intel-Oki	June 1984	Oki will provide CMOS versions of popular Intel products
American Microsystems Inc.-Hitachi	May 1984	AMI will second-source Hitachi's line of codec circuits

Table 14: US & Japan Semiconductor Agreements (1984-5) *cont.*

Companies	Date	Technology Exchanged
Motorola-Toshiba	April 1984	Toshiba will second-source Motorola's C-Quam AM stereo decoder IC
Zilog Inc.-NEC Ltd.	March 1984	Zilog will second-source NEC's proprietary V-series microprocessors and peripheral controllers

Source: Electronics Week, 6 May 1985, p.34

Table 15: Various Companies and Types of Arrangements Involved in the National and International Collaboration of Capital within the Electronics Industry

TEAMING UP TO OFFER ONE-STOP SHOPPING

In the past, a manufacturer typically supplied only one slice of the information processing pie software, communications gear, or hardware — microcomputers, minicomputers, or mainframes. Now these products are converging, and customers want to buy the entire system from one supplier. This is forcing the leading US and European manufacturers to broaden their product lines rapidly — by investing in other companies or by acquiring technology and products from them.

	Peripherals/ Components	Small Computers	Medium Computers	Large Computers	Software	Communi-cations
AT&T	Telectron (1)	In house, Convergent Technologies (4) Olivetti (2,8)	In house	No plans	In house, Intel (5) Zilog (5) Motorola (5) Digital Re-search (7) others	In house, Philips (3,8) Gold Star (3)
Bull	Trilogy Systems (2,5) Magnetic Peripherals (2)	In house, Fortune Sys-tems (2,6,8)	In house, Convergent Technologies (4), Ridge Computers (5,8)	In house, NEC (5,8) Honeywell (6)	In house	In house

Burroughs	Memorex (1) Peripheral Components (2), Qume (4), Canon (4), Intel (9)	Convergent Technologies (4)	In house, Graphics Technology (1)	In house	In house, Midwest Systems Group (1), Graphics Technology (1), others	In house, Systems Research (1)
Control Data	Centronics (2), Magnetic Peripherals (2), Trilogy Systems (2,5)	In house, Columbia Data Products (4)	In house	In house, Microelectronics & Computer Technology (5)	In house, Chrysler Corp (5), Northrop Electronics (7)	The Source (2), United Telecommunications (2)
DEC	In house, Trilogy Systems (2,5)	In house	In house	In house, Microelectronics & Computer Technology (5)	Third-party agreements	Northern Telecom (5), Xerox (5), Voice Mail Int'l (8)
Honeywell	Magnetic Peripherals (2), Synertek (1)	In house, Columbia Data Products (4)	In house, Bull (2,6)	In house, Microelectronics & Computer Technology (5), NEC (5,6,8,9)	Third-party agreements	Action Communication Systems (1), L M Ericsson (3,5,8), Keycom (3)
ICL	In house, Fujitsu (4)	In house, Logica (4), PERQ Systems (5,9), RAIR (8,9)	In house	In house, Fujitsu (5,8)	In house, third-party agreements	In house, AT&T (6,8), Mitel (8)
IBM	In house, Intel (2)	In house	In house	In house	Microsoft (4), Comshare (8), others (4,7,8)	Rolm (2), Merrill Lynch (3), SBS (2), Sears & CBS (3)
NCR	In house, Magnetic Peripherals (2)	In house, Convergent Technologies (4)	In house	In house, Microelectronics & Computer Technology (5)	In house, third-party agreements	Comtern (1), Ztel (2), Intel (8)

Table 15 cont.

	Peripherals/ Components	Small Computers	Medium Computers	Large Computers	Software	Communi-cations
Nixdorf	In house, LSI Logic (4)	In house	Spartacus Computers (6)	In house, Auragen Systems (5)	In house, Spartacus Computers (6)	In house
Olivetti	In house, Hermes Precisa Int'l (1), Lee Data (2,8), Ithaca (2,8)	In house Corona (2,8) Kyocera (4)	In house, Stratus Computer (2,8) AT&T (8)	IPL (2,8) Hitachi (8)	In house, Digital Research (2,8), Shared Financial Systems (2,8)	In house, AT&T (8), Northern Telecom (8,9), Bolt Beranek Newman (8)
Siemens	In house, IBM (4), Furukawa (3), Intel (4), Xerox (6,8)	In house	In house	Fujitsu (8)	In house	In house, Corning Glass (3)
Sperry	Magnetic Peripherals (2), Trilogy Systems (2,5)	Mitsubishi (7)	In house	In house, Microelectronics & Computer Technology (5), Mitsubishi (7)	In house, third-party agreements	In house, Northern Telecom (7)

Source: Business Week, 16 July 1984, p.54

(1) Acquisition (2) Equity position (3) Joint Venture (4) *OEM* agreement (5) Technology development (6) Technology exchange or licensing (7) Joint product development (8) Marketing agreement (9) Manufacturing agreement

100 strategic alliances with different companies in an attempt to improve its technical and market position in computers.[39] For the electronics industry the number of companies involved is impressive. In 1983 alone 146 mergers and acquisitions took place including Westinghouse's take-over of Unimation, the leading robot company (*Business Week,* 16 July 1984).[40] Also striking is the truly global character of the process of concentration as electronics and major non-electronics companies enter into national and international arrangements to strengthen the resource-base necessary for successful global competition. Out of this need has come the major acquisitive strategy implemented

by the huge US corporations Exxon, GM and GE. We also saw earlier how with government and military support, US companies were joining in industry-university R&D centres aimed at improving the United States' competitive strength in international markets, most particularly against Japan. It is now evident, however, that the struggle for control of the electronics infrastructure is not being waged on a purely national basis with governments seeking to support the electronics base of their respective countries.[41] Indeed, looking at many of the arrangements in the tables above, it seems plausible to argue that as far as corporate capital is concerned there is a pronounced trend towards the formation of international partnerships. In this respect, we have seen for instance AT&T's steps to join forces with Japanese companies to confront IBM. The same objective is behind AT&T's acquisition of a 25% share of Olivetti (Italy) (Foremski, 1984) and of the deal with Philips, the Dutch electronics giant known for its market strength, one of the major weak spots of AT&T's challenge to IBM (*Financial Times,* 17 January 1983).

In turn, IBM's strategy has been to join forces with communications and software companies and one of the most important microprocessors company in the field. In addition, it has also entered into agreement with the Japanese company Matsushita in the area of information systems. Large Japanese companies are all involved in collaboration with both US and European companies. Fujitsu, for instance, has an agreement with ICL (UK), Siemens (Germany), Intel (US) and has taken control (49.5%) of the US computer company Amdahl Corp. to enhance its own computer marketing in the US (*Business Week,* 16 July 1984). NEC has made agreements with Bull (France), Honeywell, (US) and Zilog (US), Zilog being one of the companies acquired by Exxon. Hitachi has agreements with Olivetti, Motorola and American Microsystems a company where two major non-electronics manufacturers, one German and another American, had recently become shareholders (see Table 12). Fanuc has Fujitsu and Siemens as main shareholders and in 1982 launched a joint venture with GM, GMF Robotics Corp., which has now become the leader of the US robot market (see Figure 38). More recently Toshiba entered into an agreement with AT&T to market a new AT&T switching system in Japan (*Financial Times,* 23 January 1986) and, in 1987, signed a significant technology and trade agreement with Motorola. In turn, European companies such as Philips and Siemens have collaborated,[42] and have made separate agreements with Japanese and US companies. The most publicised recent European-US interlinking has been the acquisition of ITT's telecommunications operations by the French company CGE, parent of Alcatel-Thomson (see Figure 34 above). Finally, two new forms of alliance are attempting to bring together companies from three or more countries. The first form is the agreed link between Honeywell (US), NEC (Japan) and Bull (France), the first attempt at a worldwide joint venture between major US, European and Japanese computer companies (*Financial Times,* 25 September 1986 and 31 March 1987). The second form is represented by the formation of the computer concern Counterpoint, designed to function as the keystone in a global alliance of corporate giants who do not compete with each other. Originally backed by AT&T, Kyocera (Japan) and British & Commonwealth Shipping, membership is being chosen so that their particular strength can be shared by Counterpoint (*Electronics,* 27 January 1986 and 22 January 1987). Collaboration agreements therefore are crossing all types of boundaries as companies enter into a variety of relations at national and international levels in their effort to gain the resource-requirements imposed by global competition. For instance, US semiconductors companies are joining to carry out R&D programmes to strengthen the US's stand against Japanese competition. More recently, they have entered into production and technological agreements to try to secure larger shares of the market.[43] They are also entering into ventures with computer and telecommunications companies seeking integration; and, finally, they are joining forces with foreign companies, most recently with the Japanese (see Table 14), to exchange expertise and complement their actions in areas of mutual interest.[44] The US company Intel is a good illustration of this process. Nationally, it has IBM as a shareholder but also has a manufacturing agreement with TI and Burroughs, a marketing agreement with NCR and a software agreement with AT&T. Internationally, it has agreements with three Japanese companies, namely, Oki, Toshiba and Fujitsu, the latter two themselves major semiconductor producers.

The result is that electronics companies competing in the global and convergent electronics market are not driven by an overriding national interest. Instead, they are pursuing the overriding interest of capital, that is, profits and accumulation in a context of global competition. It is this interest which, ultimately, develops into the national interest as the role of the electronics technical base assumes greater strategic importance, particularly, for the complex of social interests controlling the development of capitalist societies. Yet, for US companies competing in the computer market, for instance, their interest is not in IBM gaining strength. It is in their own share of the market. It is this logic that has led the long-established US-based mainframe manufacturers Burroughs and Sperry to merge into the new computer company Unisys. But if this further means collaborating with Japanese or other foreign companies to compete with IBM, they will tend to do so. AT&T is a good example of this attitude, but so are Honeywell,[45] TI,[46] and all the other companies we have seen above. Thus, the process of capital concentration in the electronics industry becomes international, though most companies within each team have retained their corporate identity.

It has been predicted that as few as forty highly integrated major electronics firms may come to dominate the electronics industry. To that, the following more detailed prediction must now be added

> The industry will evolve down to three sectors: a few gigantic, vertically integrated suppliers, such as IBM and AT&T, and perhaps one or two Japanese companies offering soup-to-nuts product lines and low cost, high-volume hardware manufacturing; a second, much larger tier of systems integrators assembling products from various manufacturers to create customised computer systems for different industry niche markets; and, finally, a horde of small, specialized suppliers providing the systems integrators with individual pieces of hardware or software tailored to specific markets (*Business Week,* 16 July 1984, p.50).

Whether or not, and in what way this scenario will materialise is an open question. The final outcome will be the result not only of all the struggles and alliances which accompany the restructuring of the electronics industry but, indeed, of all the processes which we have discussed above involving governments, the military and, generally, the entire complex of social interests dominating and shaping the development of microtechnology. Furthermore, the extent to which these interests will continue to shape the technology depends upon society as a whole, i.e. on the balance between all the conflicting interests, within particular social, economic, political and cultural conditions.

There are fears, for instance, that the US microelectronics capability may be weakened by internal conflicts within the social constituency of microtechnology. Military pressures and international politics are apparently conflicting with the demands generated by international economic competition, especially in the face of a challenger like Japan whose social constituency has been consistently geared to economic success. So far, however, the evidence for this is not conclusive. Moreover, the competitive struggle for global control of microtechnology has been shown as a process of greater complexity than the image of competing nations implies. First, from a technical point of view, the US's electronics industry is not being challenged with uniform intensity or success on every technological front. Second, in an industry where the commercial market is expected to reach $520 billion by 1990, the nationalistic military constituent no longer possesses the influence of the early postwar years. Third, and most importantly, electronics corporate capital is not operating on a nationalistic basis but on a global basis. This has led increasingly to international alliances as the basic resource requirements now are beyond even the largest corporations in the field. In the global arena, therefore, companies are lining up internationally, giving the United States' microtechnological process an inseparable international element. This not only satisfies the need for global markets and productive sites but allows companies rapidly to accumulate the basic resources demanded by worldwide competition. It is the interaction of these trends and all the others examined in this chapter that will determine the outcome of the current process of social shaping of microtechnology.

CHAPTER SIX

FOUNDATIONS OF THE FUTURE I

The development and diffusion of microtechnology is central to the microelectronics revolution. This technology has gradually spread into most areas of human activity and has the potential to affect society's entire technical realm.[1] The microrevolution seems inexorable. It would appear that societies are faced with an external force, a juggernaut before which they have to yield and adapt.

Yet, what the preceding chapters have shown is a completely different picture suggesting that the technological process involved in the microrevolution has no life of its own. On the contrary, at every stage it has been embedded in and shaped by sociohistorical factors. In particular, it has been shown that behind the development of the technology there is a clear social constituency, a power complex whose interests dominate the development of the technology. In the US, in the post-Second World War period, such a social constituency has been dominated by the overriding and mostly complementary interests of corporate capital, government, the military and science. In consequence, accumulation of power — economic, political, military, scientific and technological power — has been the driving force behind the technology and not any supposed benefit for mankind. Indeed, we have seen how, for all these social forces, microtechnology represents a key factor in the development of their power and interests. There is little hope that this technology will be shared, for instance, to help to alleviate Third World problems. But it does not have to be like this. In 1980, the fiftieth anniversary issue of *Electronics* sounded this hopeful note for the future

> Electronics in the form of dedicated intelligence will be every-

> where — in home entertainment, appliances, automobiles, airports, highways, offices, banks, stores, factories, hospitals, schools, and, of course, in space. Prodigious strides have already been made along these lines, but the sheer volume of information processed, stored and transmitted in the future will be mind-boggling . . . The world, then is entering . . . the 'era of computational plenty'. But to exploit this new wealth, further advances will have to be made (*Electronics*, 1980, p.448).

Also, in 1980, in his *World Challenge*, the French political author Servan-Schreiber confidently argued that

> telecommunications, microprocessors and their tendency to converge in the new creative process should be placed freely and completely at the disposal of Third World people . . . *so that they can become creators themselves* (Servan-Schreiber, 1980, p.268).

Undoubtedly, these are attractive visions. Few would deny that electronics has much to offer to mankind, yet I find it difficult to share the optimism of Servan-Schreiber or the editors of *Electronics* in my vision for the future. More specifically, it seems that to exploit this new wealth for the benefit of mankind and realise the 'era of computational plenty' for all nations, more will be needed than simply further technical advances. Ultimately, the social framework nurturing the technological revolution may need profound changes.

This view is supported by current trends dominating microtechnology as examined in previous chapters. Nowhere among these trends did we find clear indications that humane concerns such as health and the environment, improved standards of living and conditions of work, eradication of poverty, etc., are significant factors behind microtechnology's development. Granted, socially useful products are being produced and the weakening of the United States' power complex in the 1970s showed that concerns for the public good may become quite important given the right combination of circumstances. At present, however, the evidence from the United States indicates the absence of such conditions, and as a result humane concerns carry limited weight in the overall shaping of microtechnology. Paradoxically, this is underlined by those very groups who, by developing alternative forms of microtechnology, try to enhance the quality of life and work and seek to make public concern a priority. Although I shall not deal with the particulars of these efforts here,[2] it remains true that, for all their worth, their practical influence in technology development has been marginal. In the present context, perhaps the greatest contribution of these efforts lies in the creation of a valuable technical stock offering potential for a world of different and humane possibilities. From this angle, they show that there is nothing inevitable about the shape of microtechnology and, indeed, the general development of technical systems.

In practice, it has been primarily the combined interests of capital, military, government and science which have shaped and continue to shape the overall development of microtechnology and, more generally, the United States' research and development system. In competition with these social forces' need to reproduce and advance themselves, more humane concerns are relegated to second place. The most optimistic visions for the future tend to be no more than dreams based on technical possibilities but entangled in the nightmarish web of interests, control and domination that pervade technological development. The historical analysis of this book certainly points in this direction, and suggests that profound changes in the societal framework nurturing microtechnology would be the most effective way to realise its potential for the benefit of mankind. In particular, an alternative social constituency driven by different interests must emerge and shape the development of microtechnology into avenues consistent with the public good rather than focusing on the particular interests of individual sectors of society.

Why is the present social constituency inherently incapacitated to fulfil this purpose? Does its past record argue that one can only expect more of the same for the future? Is a radical change of direction not possible? or, has the past direction come from necessity meaning that the trend is likely to continue into the future? The answer to these questions is yes, there is a strong element of necessity which links past and future and which incapacitates the present US social constituency of microtechnology from following a new direction. The constituents are not consciously being uncaring. Rather, this behaviour stems from the dynamics and actions implicit in their overriding interests, i.e. those interests which, like profits for capital, cannot be negated in the long-term because they are an integral part of the social forces concerned. The inherent incapacity of the present social constituency to change course lies in the elements of necessity and spontaneity their overriding interests involve. This factor is also a source of power in today's complex society, given the difficulties implicit in the idea of radically altering present forms of development consciously to reorganise it in directions which are consistent with the best ideals of freedom, justice, democracy, and so on. Of course, different social constituents differ from each other in their overriding interests and we have seen how the greater the weight of the overriding interests of one particular social constituent, the greater the likelihood of these interests shaping the development of the technology. Moreover, as will be seen below, the place of each of the social constituents within the technological process is different, so that whereas the social constituency as a whole may undergo significant changes, there are social constituents which are intrinsically related to the nature of the technology and others which are not. All these are crucial points and they will become clear after a scrutiny of the present social constituency and its relation to microtechnology. This will reveal the extent of the difficulties involved in altering the present course of events, at the same time providing a more fundamental explanation of the major trends identified in previous chapters. This sort of understanding, I hope, will contribute to a more systematic approach, not only to make better sense of the development of microtechnology in the US (or any other country for that matter) but also for estimating the prospects of major changes occurring in the social framework involved in such a development.

A basic premise of the following discussion is that the development of technical systems always involves the integration of human, financial, material, time and space resources. The control of these resources is thus crucial since those who possess them can use them to shape technical systems in accordance with their interests. In the case of the US's R&D system and microtechnology, this control has been predominantly in the hands of capital, the military, government and science. These forces have been the dominant social constituents of US R&D and microtechnology.

Of course, it would be wrong to say that the development of a technological process is the result of the unopposed actions and interests of its dominant social constituency. The reality is much more complex and a better picture would have to acknowledge that all technological processes are, in fact, the result of the struggles, alliances and confrontations between all those forces and interests concerned with and affected by their development. Thus, in the case of the US, other social forces such as labour and the trade union movement have undoubtedly played a part in the development of microtechnology. The difference is, however, that such a role has been a subordinate one since these social forces have had nothing like the command over basic resources enjoyed by the dominant social constituents. This fundamental inequality explains, for instance, why the struggle of labour and trade unions to shape technology has been largely defensive, limited to influencing technological initiatives emerging from the dominant social constituency. From an analytical point of view, it also justifies my decision to concentrate primarily on understanding the dominant social constituency as the best way of unveiling the social shaping of microtechnology.

Dissecting the Dominant Social Constituency of Microtechnology

Previous chapters have shown that science, capital, the military, and government are all members of the dominant complex of interests shaping the development of US microtechnology and R&D system. We saw the relative weights of these constituents changing throughout the postwar period as they interacted with the technology's development and changing historical circumstances. We also learned that not all the dominant constituents seem to be necessary for a country to possess strong capabilities, as the case of Japan consistently showed in relation to the military. To be more precise, it should be added that the military can be a necessary dominant social constituent when, as in the US, a country's development in microtechnology is strongly biased toward military purposes. But this only reasserts the point that the direction of technological development is invariably the result of the interplay of dominant interests among its social constituency.

The important issue I am about to examine is how each of the present social constituents relates to the nature and development of microtechnology. That is, it must be determined whether they are equally important, or whether there are social constituents whose displacement would have a profound impact on the development of microtechnology and its societal framework. In this sense, the point is to determine whether the displacement of some constituents would imply a radical alteration in the nature of society or, even further, a fundamental threat to the very existence of microtechnology. An insight into these issues will help us to assess the magnitude of the tasks necessary for transforming the present social constituency and hence, for any attempt to alter its course of development radically in pursuits of humane concerns. In practice, we shall find that some trends will prove much more difficult to eradicate than others, in the same way as the development of radically different trends may prove to be a major social challenge.

a) The Science Constituent

The science social constituent is the only *dominant* constituent which can be considered as rooted in microtechnology[3]. It is what I call an intrinsic social constituent of microtechnology, meaning that its absence from the social constituency implies a fundamental block to the very existence or further development of microtechnology. The reason for this lies in the fact that microtechnology is a science-related technology,[4] i.e., a technology where scientific knowledge and, more generally, R&D, are basic resources inseparably related to its development. For instance, had it not been for the scientific understanding of crucial properties of matter and energy, microelectronics and hence, the microelectronics revolution would simply not have developed under any societal condition. A closer look at the relation between science and microelectronics will support this statement.

Both electromagnetism and quantum theory are generally acknowledged as having laid the foundations for the present-day microelectronics revolution.[5] In Atherton's words,

> From the discovery of electromagnetism it is possible to trace a continuous development of understanding spanning more than a century that incorporates the electromagnetic theory of light, the beginning of relativity, and quantum theory and quantum mechanics. From the latter came our understanding of semiconductors and the . . . silicon chip (Atherton, p.14).

Electromagnetism and the advances made in the nineteenth century had opened the way for electronic telecommunications as well as for the generation of electricity. The discovery of the Edison effect in 1883[6] established the basis for the development of the vacuum diode by John Fleming in 1904 and the vacuum triode by Lee de Forest in 1906. Both elements, the diode as detector and the triode as amplifier and switch, became the elementary active components of all electronics (e.g., telecommunications devices and computers) up to the advent of semiconductor devices.

The scientific base of solid-state electronics is a well-documented fact, particularly in relation to the epoch-making discovery of the transistor.[7] The roots of its invention are often traced to Faraday's discovery, in 1833, that the conductivity of silver sulphide increases with temperature (while metals display an opposite effect) and to the observation of other phenomena which in the nineteenth century puzzled scientists[8] and were explained only with the development of quantum theory in the twentieth century. Thus, in

1932, A.H. Wilson published his work on the quantum theory of semiconductors which built on the work of other scientists like F. Bloch and A. Sommerfield.[9] By 1933, as Wilson claimed, 'all the basic principles concerning the solid state had been established' (quoted by MacDonald *et al.*, 1981, p.177).

The transistor, however, was developed only in 1948, ushering in the era of solid-state electronics with its huge expansion and integration of electronic systems. That its development took fifteen years shows that scientific knowledge does not lead automatically to technology. Indeed, social and technical factors such as the need for equipment to control the level of impurities in semiconductor materials and, as we have seen, the spur of commercial or military interests played a crucial role in the transistor's appearance. Without science, however, there would have been no transistor. As Braun and MacDonald have so forcefully stated,

> More than most innovations, the transistor was born out of scientific discovery. No doubt the science was aided by a whole gamut of techniques and instruments, but these served as the tools of science. Many innovations are based on technology, often aided by science at many stages. The transistor is one of the supreme examples of an invention truly based on science (Braun and MacDonald, 1977, p.167).

In the transistor, therefore, science was an integral part of the technological process, in the same way as the science of electromagnetism had been the base leading to early developments in telecommunications. Today, we are in the midst of the microelectronics revolution and nothing has changed regarding the importance of scientific activity and, more generally, R&D. Indeed, the relentless advance in scientific and technical knowledge continues. Science and R&D are as much an inseparable basic resource of microtechnology as they ever were and, most likely, ever will be. Ultimately, this is where our science social constituent becomes an intrinsic constituent of microtechnology; it is its double character as a social force and a basic resource of microtechnology which explains its essential importance. Of course, we have seen that the pursuit and technical exploitation of scientific knowledge takes place under socially determined conditions, which means that science is itself shaped by the wide interplay of social interests.[10] This fact, however, does not affect the extent to which microtechnology is science-based.

b) The Capital Constituent

Unlike science, the capital social constituent is not rooted in the technical nature of microtechnology. Nevertheless, microtechnology cannot develop in the abstract but is a sociotechnical system where the technical interacts with the social, shaping and being shaped by the societal context. Capital can thus be seen as another intrinsic social constituent of microtechnology development. In a capitalist society such as the US, the socioeconomic system is generally dominated by capital. As a social force this ultimately embodies capitalism, realising it through its actions, rationality, driving forces and social pursuits. Thus, when a sociotechnical system such as microtechnology develops within such a context, i.e. as a capitalist sociotechnical system, then capital is bound to be an intrinsic social constituent of such a system. By definition a capitalist sociotechnical system exists only when the overriding interests of capital fundamentally shape its development.

We know already that capital is one of the dominant social constituents of the United States' microtechnology and R&D system and, indeed, the one currently carrying the greatest relative weight within both sociotechnical systems. Various crucial features and trends observed in the analysis of both systems (e.g. direction and pace of technical change, concentration of capital) find their roots in the rationality and interests of capital. This is so because, in its role as intrinsic constituent, capital carries with it imperatives and trends which strongly influence the shape and dynamic of development of sociotechnical systems. Closely examining the nature of capital reveals the most important mechanisms and tendencies involved.

A starting point is given by Kaplinsky, who states that capital

> can be defined as the constellation of forces which owns machinery, buys in labour and other inputs, organizes production and marketing, is responsible for production and process development, liaises with the state and copes with competition from other firms. As such in order to operate effectively, it is essential that it retains control over social, technological and economic relations of production. So for capital, the role of technology is not only to produce attractive commodities at low cost, but also . . . to allow control to be exercised within the factory itself. (Kaplinsky, 1984, p.111.)

The questions which naturally follow are why does capital

operate in this way, and what are the basic trends resulting from its operation? The answer lies in the 'restless never-ending process of profit-making' (Marx, 1867, p.151) which underlies capital and the constant dynamic of capital accumulation associated with it.[11] 'Accumulate, accumulate! That is Moses and the prophets!' (*ibid.* p.558), wrote Marx, highlighting the fact that under capitalism, social production becomes dominated by accumulation for its own sake. This means that a portion of the profit is recurrently converted into additional capital, thus ensuring its constant self-expansion.[12] In Sweezy's words, this process 'constitutes the driving force of capitalist development' (Sweezy, 1970, p.80).

A natural result of the process of capital accumulation is the concentration of capital — that is, the growth in the scale of production commanded by units of capital (e.g. individual corporations) in line with their own growth or accumulation. But according to Marx (1867), there is another powerful tendency of capital which leads to the increase in the scale of production and, indeed, to its agglomeration in huge units. This is the process of centralisation of capital which brings together units of capital which already exist, thus transforming many units into fewer large ones.[13] The underlying factors in centralisation are competition and the associated importance of economies of scale [Mandel (1977), Sweezy (1970)]. The latter phenomenon, which in the market expresses itself in the existence of ever-higher 'entry barriers' [i.e., 'factors which make it difficult for a new firm to enter the market' (Kaplinsky, 1984, p.112)],[14] tends to be self-reinforcing in that, as he explains, once the

> scale economies exist there will be an inevitable tendency towards the concentration of production in large-scale, technically efficient plants, leading to the centralization of production in particular areas and particular firms (Kaplinsky, p.114).

In this context, some of the most crucial trends in concentration and centralisation of capital are as follows. First, the progressive integration of isolated processes of production into 'processes of production socially combined and scientifically arranged' (Marx, 1867, p. 588).[15] Secondly, the increasing pace of technical change as a result of the expansion in accumulation and vice-versa. This process manifests itself in the incorporation of labour-saving techniques and leads to a constant increase in the relative participation of capital in the production process.[16] Thirdly, the progressive replacement of competition among a large number of producers by monopolistic or oligopolistic control over markets by a small number of giant units of capital (Sweezy, 1970).[17]

In one form or another, these factors have been at work throughout the entire period of our concern, underlying the development of microtechnology within the framework of US capitalism, and inextricably tied to the role of capital as an intrinsic social constituent of US microtechnology. Their presence and influence can clearly only be affected by altering the role of capital within the social constituency of the technology. Within the broader context of capitalism, however, this is not an easy goal, given the pervasive part played by capital in the control and development of socio-economic and technological processes. Indeed, it is difficult to imagine a radical alteration of the role of capital in the social constituency of microtechnology, without a similar radical alteration in the broader societal context of capitalism. This implies that, in the same way that the science constituent was central to the very existence of microtechnology, the capital social constituent is central to the very existence of the capitalist societal framework within which it develops. Consequently, the reorientation of microtechnology away from the trends currently imposed by capital seems likely to involve revolutionary social changes, particularly if this reorientation is towards humane goals.

c) The Government and Military Constituents

Thus far, we have dealt with government and the military as separate social forces. In practice, however, government and the military interact with each other in the development of the state. The military, for instance, obtains most of its financial resources from government and, to the extent that military power is regarded as a necessary component of political power, interests tend to coincide. In the US, this is reflected in government largesse towards the military and the considerable influence of the military itself within government. Government interests, however, are much broader than those specific to the military alone, thus justifying treating them as separate social forces. Tension can develop between these forces, with the government occasionally restraining military demands, as happened in the aftermath of the Vietnam War.

The importance of dealing with government and the military as separate social forces becomes even more apparent as we look at microtechnology. In my view, while the military is not an intrinsic social constituent of this technological process, government certainly is. Thus, in the case of the military, we know already from the Japanese experience that this social force is not necessary for the development of strong capabilities in microtechnology. The military is therefore what I prefer to call a non-intrinsic social constituent, i.e. a constituent which, whatever its influence in shaping the development of microtechnology, is not necessary, or does not play an essential part in its development. In practice, the main role of a non-intrinsic social constituent is that its interests may contribute to determine the shape of a sociotechnical system such as microtechnology. Often this helps to explain differences in the development of similar technological processes in different countries, as, for example, the difference between the US and Japanese microtechnology. The presence of strong military interests within the US social constituency has provided strong militaristic overtones in the development of this technology while, in contrast, in the Japanese social constituency the role of military interests has been conspicuous by its absence.

The role of government in the social constituency of microtechnology is altogether different from that of the military. The reason for this lies, primarily, in the technological nature of the microtechnological process and the large scale of human, material, financial, and time resources involved. We have seen how the process of convergence of technologies and industries is advancing rapidly, with the result that a process of intra-national and inter-national collaboration of capitals is underway. This process has created the need for government to develop into an intrinsic social constituency of microtechnology, given that the resources required by far exceed even the ability of large corporate capital to satisfy them. In addition, it is also necessary to consider the direct role of governments in technological policy-making and hence in the creation and alteration of the framework of legislation shaping the sociotechnical system. In this respect, one may consider government institutions such as policy-making bodies as indirect depositories of basic resources for microtechnology, since they possess regulatory powers and hence, the ability to influence the availability of resources through taxation, price-regulation, royalty and tariff policies, etc. Added to this are government institutions such as R&D laboratories and public companies which are, in fact, direct depositories of microtechnology's basic resources. Thus there are good reasons for putting government within the ranks of the intrinsic social constituency, although one has to make clear that, unlike science or labour, its significance is not directly rooted at the level of basic resources but, rather like capital, at the level of being an institutional depository of basic resources within a capitalist society.

So, what happens if government and the military are displaced from the social constituency of microtechnology? As regards government, the answer points in a similar direction to that of capital's role in the social constituency of microtechnology; that is, major structural changes in society will have to take place for government to become unnecessary to microtechnology. Clearly, the whole thrust of current trends will have to be altered, with a considerable reduction in the current centralisation, scale and pace of microtechnology development. Only in a situation like this can one envisage resource requirements being met by social constituents other than government. For such a situation to materialise, however, there must be radical changes within the societal structure shaping microtechnology.

In this light, it seems that, rather than seeking the displacement of government from the social constituency of microtechnology, a more realistic alternative would be to change the role of government within this social constituency. The potential of this alternative was seen in the aftermath of the Vietnam War, when the US government was much more supportive of technological developments involving greater social concern. Of course, at that time the US government was reflecting, on the one hand, the relative weakening of the wartime social constituency and, on the other, the emergence of public opinion which, similar to other social constituents, was an influence in moving the technological process into new avenues of social concern. It is thus not government alone that counts, for as a social constituent of microtechnology government is, simultaneously, a battleground where the interests of all other social forces meet in the political struggle to shape the development of technology. In this respect, it is the interrelations of government with other social constituents which become crucial since it is here that its role in shaping microtechnology is determined.

In practice, government shows few of the trends which explain capital's behaviour. Rather, insofar as its own interests become deeply associated with the interests of capital — or those of other social constituents such as the military — government tends to adopt those trends and imperatives as its own. A power complex thus takes shape from this interrelation between the dominant social constituents which makes radical change a much more difficult affair than it may appear when looking at each of the social constituents separately.

This becomes clear when one considers the possible displacement of the military. In isolation, the displacement of the military from the social constituency of microtechnology would neither threaten the existence of microtechnology's capabilities nor demand fundamental changes in US society; indeed, as already seen, some analysts are arguing that it is the heavy influence of the military constituent which in the context of strong international competition threatens US microtechnological capabilities. In terms of microtechnology development, therefore, the military seems a softer target to attack to alter the dominant social constituency of US microtechnology and hence, its development. As a non-intrinsic social constituent, this is quite true, but, for the same reason, it is also true that the displacement of the military constituent alone will not ensure the emergence of a new social concern within the process of microtechnology. Rather, it may ensure the reduction of the heavy militaristic purposes currently influencing this process, while leaving in control the trends propelled by the need to accumulate capital. Obviously, this displacement in militaristic purposes is a valuable step in the direction of humane concerns and, indeed, the one most likely to be successful, given the present situation of conflict in the simultaneous pursuit of strong militaristic and commercial goals in microtechnology. Any advance in this direction, however, demands the decisive participation of government and, ultimately, of the social forces influencing decision-making. This brings us back to the issue of interrelations between the dominant social constituents, and hence to the power complex shaping the development of US microtechnology.

Some Conceptualisations of the US's Social Complex of Power

In the Appendix, the historical roots of the capital-government-military-science power complex are traced back to the attempt to marshal science and the infant R&D system into the military effort demanded by the First World War. It was during the Second World War, however, as this complex was reactivated to direct the war effort and, specifically, the scientific and military effort, that momentous changes began to occur. In comparison with the First War, the scale of the Second World War's scientific and technical effort, and its lasting impact on the development of the R&D system, and indeed on society as a whole, represented a considerable leap forward. This time, as we have seen, US society found itself with a science-and-technology-based social complex of interests which not only outlasted the war effort, but has come effectively to dominate the country's political and economic life. The impact of this development has been the subject of much analysis and interpretation.

Various scholars have identified the Second World War, and

particularly the Manhattan Project,[18] as the critical point in the emergence (or reconstitution) and development of the complex of social interests and power. The nature of such a complex has been the subject of different theories. Some of the most influential have been proposed by scholars such as L. Mumford, J. Galbraith and the theorists of the military-industrial(-scientific) complex (MIC), particularly within the Marxist tradition.

For instance, Mumford's Pentagon of Power controlling the modern Megamachine is the ensemble of social interests formed by the military-industrial-scientific élite plus the bureaucratic and educational establishments. This Megamachine emerged under the pressure of the Second World War, but, following the war, according to Mumford,

> it did not give up its absolute weapons or the scheme for universal domination by threat of total destruction that had given a coalition of scientific and military agencies such inordinate power. Far from it. Though nominally the older organs of industry and government resumed their diverse activities, the militarized 'elite' fortified themselves in an inner citadel . . . cut off from inspection or control by the rest of the community. With the pusillanimous aid of Congress, they extended their tentacles throughout the industrial and academic world, through fat subsidies for 'research and development', that is, for weapons expansion, which made these once-independent institutions willing accomplices in the whole totalitarian process . . . [thus] . . . In a short time, the original military-industrial-scientific elite became the supreme Pentagon of Power, for it incorporated likewise both the bureaucratic and the educational establishments (Mumford, 1970, pp.226 and 269).

In turn, J. Galbraith's view (1974, 1978) of the US's complex of power is one that emphasises how big corporations — governed by the technostructure — and the state are now deeply enmeshed in a network of common interests and goals. Thus,

> No sharp line separates government from the private firm; the line becomes very indistinct and even imaginary. Each organization is important to the other; members are intermingled in daily work; each organization comes to accept the other's goals; each adapts the goals of the other to its own . . . [thus] . . . The state is strongly concerned with the stability of the economy. And with its expansion or

> growth. And with education. And with technical and scientific advance. And most notably, with *the* national defence. These are the national goals; . . .[whereas] . . . The technostructure requires stability in demand for its planning. Growth brings promotion and prestige. It requires trained manpower. It needs government underwriting of research and development. Military and other technical procurement support its most developed forms of planning. At each point the government has goals with which the technostructure can identify itself. Or, plausibly, these goals reflect adaptation of public goals to the goals of the technostructure (Galbraith, 1978, pp.309).

But, it is with the military that big corporations find the best conditions for their goals.

> The Department of Defense supports . . . the most highly developed planning in the industrial system. It provides contracts of long duration calling for large investments of capital in areas of advanced technology. . . This leads the technostructure to identify itself clearly with the goals of the Armed Services (*ibid.*, p.310).

Finally, the interlinking of the scientific and educational estate within the above complex of social interests takes place through two interrelated processes. First,

> the educational and scientific estate is no longer small. . . it is very large. It is no longer dependent on private income and wealth for its support; most of its sustenance is provided by the state. . . [Second]. . . the technostructure has become deeply dependent on the educational and scientific estate for its supply of trained manpower. It needs also to maintain a close relation with the scientific sector of this estate to ensure that it is safely abreast of scientific and technological innovations (*ibid.*, pp.289-90).

Another author who has written on similar lines to Galbraith is D. Price. See *The Scientific Estate* (1965).

Finally, the military-industrial(-scientific) complex has been defined by Pavitt and Worboys 'as a coalition of certain industrial interests, the military, big science and technology, and others who profit from the proliferation of war and have an interest in preparations for such eventuality' (Pavitt and Worboys, 1977, p.26). Various explanations have been given for the reasons behind the MIC. Reich, for instance, argues that the

> growth and persistence of a high level of military expenditure is a natural outcome in an advanced capitalist society that both suffers from the problem of inadequate private aggregate demand and plays a leading role in the preservation and expansion of the international capitalist system (Reich, 1977, p.296).

This explanation follows Sweezy and Baran's argument that monopoly capitalism (i.e. capitalism dominated by giant corporations) needs a military machine in order to absorb the surplus and to confront the socialist system. The authors state that 'the need of the American oligarchy for a large and growing military machine is a logical corollary of its purpose to contain, compress, and eventually destroy the rival world socialist system' (Baran and Sweezy, 1975, p.190). See also Soukup (1976). The crucial economic role of the MIC is also emphasised by Mandel. For him,

> Armaments economy, war economy, represent the essential replacement markets which the capitalist system of production has found in its age of decline. . . [it]. . . is indispensable for making profitable use of the capital of heavy industry and the 'overcapitalized' big monopolies. But the arms economy makes the state the chief customer of this industry. The special ties between the state and monopoly capital. . . thus assume a more specific form (Mandel, 1977, pp.522-3).

Finally, the reproduction of the MIC has found a different explanation in Kurth's 'follow-on imperative'. In Kurth's view, military production lines are seen as national resources by all interested parties. Thus, the

> Defense Department would find it risky and even reckless to allow. . . large production lines to wither and die for lack of a large production contract. . . Such a contract renovates both the large and established. . . corporation that produces the weapons systems and the large and established military organization that deploys it (Kurth, 1972, p.308).

Clearly, these different theories agree fundamentally that the complex of social power which took shape during the Second World War has developed into a long-term, dominant feature of US society.

The Power Complex as Dominant Social Constituency of Microtechnology

In the US, the role of military and/or commercial competitive pressures has been paramount in galvanising capital, government, the military and science into an effective social complex. But it has been the power — real and perceived — wielded by microtechnology and the R&D system which has naturally attracted them into these sociotechnical systems. In this respect, we know already that, although not free from conflict, the *raison d'être* and overriding interests of each of these forces are largely complementary. In fact, for all cases we have reduced them, ultimately, to the furthering of power, in one form or another. In the case of capital this is seen in the accumulation of capital through profit-making activity; for the military in the accumulation of destructive power through improved weaponry; for science in the accumulation of scientific and technical knowledge through the advancement of the frontiers of this knowledge; and for government in the accumulation of political power both nationally and internationally, partly through the economic, military and scientific-technical power derived from the other forces.

These are the underlying purposes which have inevitably made microtechnology and R&D the focus for the convergence of capital, government, the military and science. Indeed, it can be said that it is here, in sociotechnical systems such as microtechnology and R&D, that these social forces manifest themselves as a power complex influencing the shape of society as a whole. For instance, we have seen in Chapter 3 how, in the US today, the development and use of science-based or high technology has been claimed as the way to strengthen the country's industrial and military base. The military aims at making increasingly sophisticated weapons, while industry sees in it the key to enhanced productivity and competitive power. The government is deeply involved in both processes and science must produce the results necessary for their success. In this way, they interrelate through and in the technology, thus becoming an integral part of its development, or what amounts to the same thing, its social constituency.

In the US, this social constituency controls the basic resources of science-based technology. As a result, not only it has been able to make them available for the advance of this technology but, most crucially, it has done so in such a way as to shape the development of this technology, primarily in accordance with its members' interests. In other words, each of the social constituents, while advancing science-based technology, has expressed itself through this technology, thus furthering and reproducing itself in interaction with the other constituents. As we have seen, in general, this has taken place in accordance with the relative influence of the social constituents at any given time, something that in itself has depended on the historical circumstances of that time.

In my view, the evidence from the postwar period strongly confirms that it is this social process that accounts for the shape and dynamism of the development of both US microtechnology and the R&D system. The resilience and effectiveness of the power complex is indisputable. Its role as social constituency, its ability to further technology in order to develop itself, are all facts which any search for an alternative development must acknowledge.

CHAPTER SEVEN

FOUNDATIONS OF THE FUTURE II

The emergence of alternative forms of development in microtechnology, or any other sociotechnical system, demands that the control of the basic resources of the technology shifts from the present social constituency into the hands of a new power complex whose goals lead into what we have referred to as humane paths of technological development. This means that not only is the disintegration of the present complex of power necessary, but something perhaps more difficult is needed at the same time, namely, the emergence of an alternative social constituency which both displaces the present one and, by taking control of the basic resources of the technological process, is able to infuse it with humane ideals.

At this point a number of questions arise. For instance, one wonders first how a lasting alternative social constituency raising humane ideals can emerge from present social conditions to challenge effectively the existing dominant social constituency of microtechnology and the R&D system. Can these ideals acquire the urgency of overriding interests — which has proved so successful for the present social constituency — so as to bring an element of necessity and momentum to the development of technology? And in the same vein, which social forces embody these humane ideals in their *raison d'être* in the same manner as capital embodies the purpose of profit-driven accumulation?

There are no easy answers to these questions, but my immediate reaction is that any answers will point in a pessimistic rather than in an optimistic direction. First of all, consider that after more than four decades since its reconstruction in the Second World War, the present social constituency of interests is quite strong, giving no indication of imminent collapse. Indeed, evidence suggests that with its participation in long-term programmes in microtechnology and R&D, this constituency is likely to maintain its dominant role well into the next century. Look for a moment at the position of each of the dominant social constituents of microtechnology and the R&D system. Not only does the dominant role of science, capital and government look unassailable, but the dominant role of the non-intrinsic military constituent looks quite safe despite controversy over its relative influence in shaping the development of technology. What we are implicitly admitting here is that the capitalist framework of US society is hardly likely to undergo radical transformations — this raises again the problem of finding alternative social forces — and is most likely to continue into the twenty-first century. Logically, this also means admitting that I see little possibility of a strong alternative power complex taking control of the development of US technology in the next few decades. For this to occur, the relations between the social constituents of the present power complex should already be decomposing, while at the same time new relations should be constantly emerging. At the moment, there is little indication that this will happen and, in fact, after its nadir in the aftermath of the Vietnam War, the present power complex has been gaining strength by establishing a web of institutional arrangements involving the dominant constituents of science, capital, military and government.

The Present Constituency of Power: The Path Ahead

Looking back to try to find clues for the future, it is possible to say that the closest the present social constituency of power came to disintegration was with the Vietnam War. It was during this period that Lewis Mumford wrote of the revolt of the youth against a power-centred 'civilisation '. Beneath this revolt, he stated,

> is a deep and. . . well-justified fear that the next step in technological progress may bring about the annihilation of man. With good reason the young regard the atrocious methods used in conducting American military operations in Vietnam not only as a threat against their own existence but as an ominous prelude to the whole human future (Mumford, 1970, p.372).

In this and other contemporary signs of revolt, Mumford saw the beginning of a radical transformation in the power structure of society. 'In widely scattered movements', he wrote, 'the decentralization of power has already begun. The dismantling of the entire megamachine is plainly the order of the day' (*ibid.*, p.376). Nowadays, however, when military aides to the President of the United States become popular heroes after blatantly breaking the law with covert military operations, it is clear that times have changed since Mumford's assessment. With hindsight, it is plausible to say that the difference between the Vietnam period and the present lies not in the abhorrence of wars of intervention against foreign countries and the lethal nature of military means, but in the abhorrence of high human and material losses and eventual demoralisation attached to a war of intervention which results in defeat. This is confirmed by the results of the rapid and successful US intervention in Grenada in 1983 and the very different attitude towards intervention in Nicaragua, despite the ardent wishes of the Reagan administration, owing to the clear potential for 'Vietnamisation' involved in such a military step.

Thus defeat and national demoralisation is what the dominant social constituency dreads, not wars, either 'hot' or 'cold', nor even fantasised wars such as the current Star Wars development. Indeed, history has shown that on the whole the presence of strong military galvanising forces has been invigorating for the present power complex and particularly so for the military constituent. This at least was the experience of the Second World War, the Korean War, the Cold War, the beginning of the Vietnam War and the present rearmament process under the pressure of the Soviet 'threat'. Only its costly inability to succeed has ever exposed the power complex's fundamental lack of humane purposes and has threatened its otherwise firm control of the US power structure. In my view, it is only in a critical situation like this, when the power structure ironically reveals itself powerless to influence events which inflict moral and/or material harm on the population, that the foundation of the power complex is weakened and it is possible to question and challenge its motives and control over society. Somehow people must be shocked out of their confidence in the power complex and thus recognise the irrationality of its means and ends. In contrast, at those times when the complex exudes power and achievement, pervading the population with the 'we-are-great' feeling, confidence in the complex may reach its peak, making the idea of a radical alternative look like an impossible utopia.

Today in the US, after about a decade of regaining the confidence and strength lost in the early 1970s, it is apparent that the social constituency of power is back in firm control. Indeed, it seems that most US citizens agree with the argument promoted by those in control, namely, that those years when the operation of the complex was disrupted by the rise of greater concerns for social responsibility, were in fact years of relative decline in the country's industrial and military power. Today things are different. Once again we see frequent demonstrations of US military might around the world, coupled with promises of restoring US industrial supremacy currently threatened by international competition. For anyone who lived through the critical years of the power complex, there can be little doubt how greatly it has regained the confidence of the US people.

However, this process is not yet completed and there are still clouds hovering which may spell trouble, at least for the military constituent of the complex. The development of microtechnology has shown that the galvanising forces of international military and commercial competition may no longer produce the complementary results of the immediate post-Second World War period when other countries' presence in the international market was weak. Today, when Japan and Europe are strong commercial challengers, the US social constituency of power may reveal important internal conflicts which will create tensions between the different constituents. For instance, there is strong evidence that in the face of competing social constituencies consistently geared to commercial goals, the shaping of microtechnology and the R&D system toward military purposes is an increasingly ineffective way to develop commercial technology. Indeed, it could be said that the stronger the impetus of commercial competition the less likely it is to complement strong military forces. Admittedly, I am discounting the possibility of using military strength to support commercial 'success' which, undoubtedly, would demonstrate the strongest complementarity between military and commercial undertakings. Apart from this unlikely possibility, the strong challenge of commercially focused social constituencies, like the Japanese one, may well open up cracks in the US constituency, juxtaposing military and commercial interests. Ultimately, this may lead to a reduction of both the strength of current military pressures and of the relative influence of the military constituent.

We know already that the relative weight of the military, and of the other social constituents, is not static but varies with the evolution of sociotechnical systems and changing historical circumstances. The defeat in Vietnam, for instance, plunged the military's influence to its lowest point, while recently they have staged a revival. My view is that the current trends we have been examining will eventually throw into question the viability of a strategy which gives the military control of sizeable resources of the R&D system and microtechnology. The point is that in the US social complex of interests, government, science and the military itself depend on the competitive success of the capital constituent in the world market. This means that the military's control of basic resources can extend only to the point where it begins to affect the competitive performance of the capital constituent. As we have seen, this is an issue which is being hotly debated in the US, and there is strong evidence that the military burden may be affecting US industrial and economic performance. For example, the US is now running by far the largest trade deficit in the world and we have seen that in the crucial high-tech sector it has also been rapidly losing its advantage. The US also has the largest public debt in the world, standing at over $1.8 trillion in 1985 (Dumas, 1986). Indicators of this kind cannot persist for long without, eventually, leading to a critical situation where confidence in the complex of power is again undermined unless clear corrective measures are taken. The Wall Street crash of 1987 is a clear indication that this is already happening. Alternatively, attention is diverted to other countries who are blamed for one's own difficulties and failures. Evidence of this is found in the growing feelings of resentment against the Japanese for their continuous success in the international market. Japan has been repeatedly charged with unfair trading and other practices and, for many, it would appear that Japan is the cause of US problems and not the internal contradictions of the current US strategy. This resentment recently exploded over the Toshiba Corporation's sale of computerised machine tools to the Soviet Union which would allegedly jeopardise US national security. While it is true that Toshiba broke a US-Japan agreement by doing so, the reaction in the US has been quite out of proportion, with members of the US Congress calling for the banning of all Toshiba products from the US market, around $2.3 billion worth of products (*Newsweek*, 13 July 1987). Also, in an unusual display of anger, ten members of Congress have smashed a Toshiba radio with sledgehammers on the Capital lawn (*ibid.*). This Japan-bashing, however, cannot continue indefinitely. After all, Japan is a strategic ally of the US and, as we saw in Chapter 3, international collaboration between US and Japanese companies is creating a web of interests which will make it difficult to attack Japanese interests without simultaneously hurting some of the US companies involved. Indeed, as one observer disclosed, in the case of Toshiba, 'Reportedly some American companies are discreetly rallying to Toshiba's side arguing that if Toshiba's exports to the US are banned, their companies would be damaged' (*Financial Times*, 2 July 1987, p.4).

Discounting the possibility of an all-out trade war with Japan, it is plain that corrective action must focus on the internal weaknesses and conflicts present within the US social constituency of interests. It seems clear to me that, if and when this corrective action is forced on the social complex of power, the military's control of the basic resources of technological development will be reduced, thus bringing a decline in its relative influence over microtechnology and the R&D system. There is already evidence pointing in this direction as defense growth has begun to slow down in the US.

> Program cutbacks and stretchouts will continue throughout the late 1980s as Congress tries to control the high federal deficit. This will pull the defense industries towards zero real growth. Following a brisk 9.9% gain in 1985, military equipment output is expected to climb just 3.8% this year, 1.7% next year and 0.6% in 1988 (*Semiconductor International*, September 1986, p.37).

One development which, if it continues, will help to reinforce this possibility is the change started under Gorbachev in the Soviet Union. Thus far, the indications are that those in command of the Soviet social constituency of power are seriously interested in advancing a process of disarmament which will free enormous resources now tied up in military undertakings. The Soviet Union's industrial and technological base has become increasingly weighed down beneath the military's heavy use of basic resources. In this sense, according to the trends we have seen for the US, it may well be that the Soviet Union is showing the US an image of problems to come as the US military burden continues to affect its international and technological performance. It may well be, therefore, that the Soviet's need to reduce the relative weight of its own military constituent will hasten a similar process in the US, in a context which demands and produces a weakening of the military galvanising pressures currently stimulating the military constituents in both countries.

Attaining this situation is not without problems. There are at least two points to consider. First, the importance of the web of interests and institutional arrangements which hold the social constituency together as an effective social complex. These arrangements, once they have gathered momentum, tend to be resilient to change, thus becoming a conservative force as Thomas Hughes (1984) has rightly pointed out. Nowadays in the US, for instance, as in the case of the Strategic Defense Initiative, those in control are busily creating links and relationships which will protect and nurture their project in the face of opposition. The greater the momentum of the present social constituency, the more difficult to alter its content and, hence, the development of the sociotechnical processes under its control. How far the present social constituency of power in the US has gathered momentum on the basis of military programmes is difficult to say, but it is undeniable that it has advanced rapidly in recovering the ground lost in the early 1970s.

The second point to consider is that galvanising pressures are not all external or unable to be influenced by social constituents as for instance, a geographical catastrophe would be. In the latter case, social forces are confronted with enormous pressures and must react in consequence. But there are other kinds of pressures which are not as unpredictable in nature and which, above all, may be influenced by the direct action of social forces. These include military pressures which, unless imposed by an external aggression and especially during peace time, are open to manipulation by tactics, strategies and policies which may either defuse or heighten international politico-military tension. The problem this creates is that the military have a vested interest in military galvanising pressures and tend to seek to reproduce these forces as a means of furthering their own overriding interests.

Any further development toward the reduction of the military's influence, therefore, will have to overcome the above two problems. Failure to do so makes it likely that the US social constituency of power will be responsible for the aggravation of the present trends of relative industrial and technological decline in the face of increasingly strong competition from commercially geared social constituencies. Of course, the galvanising force of competition itself is also susceptible to manipulation in a variety of ways such as tariffs, and reserved-market policies. The United States is already using this alternative to a certain extent, and there is little doubt that protectionism has grown considerably in the last few years. Full-blown protectionism, however, is a different matter since it would invite retaliation and eventually a trade war that would seriously disrupt the international capitalist system. This situation obviously does not appeal to the long-term interests of capital or the government constituent, so protectionism is more likely to be used as an instrument of 'persuasion' rather than a full-blooded tactic. Indicative of this was the $300 million of tariffs imposed on a range of Japanese electronics goods as a retaliatory measure against alleged Japanese 'dumping' of

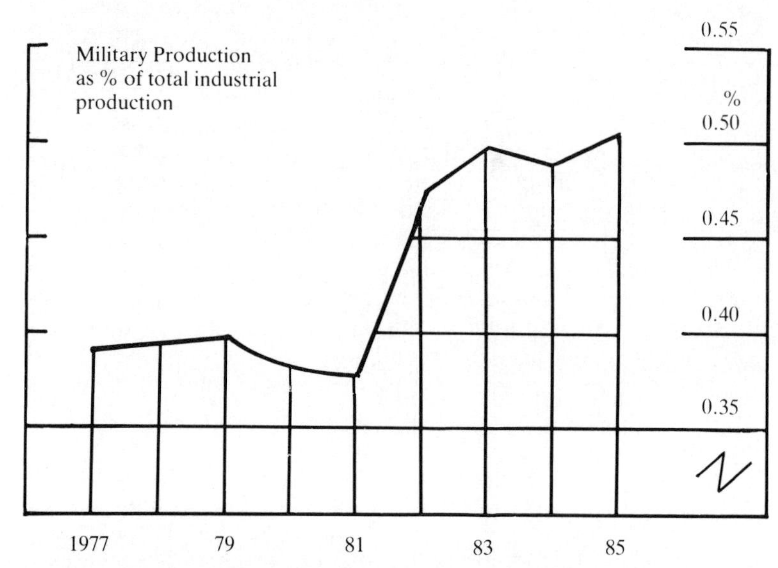

Table 40. Japan's Military Production as Percentage of Total Industrial Production (1977-85)
Source. The Economist, 13 February 1988, p.66

semiconductors in the world market. Subsequently, the tariffs were reduced by $135 million following the positive Japanese response against the sanctions (*Financial Times*, 9 June 1987). The other alternative adopted by the US social constituency is to pressurise the social constituencies of strong competitors such as Japan into high military expenditure. In this way, Japan's focus on commercial markets would be weakened and probably its competitive position would be eroded, with the result that commercial pressure on the US social constituency would be relieved without sacrificing the military constituent. Japan's long-term military strategy is not clear but an increase in defense spending and production has been taking place during the 1980s (see Figure 40), as the government accepts the argument that they must do more to share the costs of regional security with the US. The country has already breached the ceiling of 1% of gross national product it once imposed on its defence spending and, for the fiscal year 1988, it is expected to spend about $40 billion on defence, more than any other nation except the US and the Soviet Union (*The Economist*, 13 February 1988). These figures, however, must be put in the context of Japan's very small military industry which, at the moment, accounts for only 0.5% of Japan's industrial output. In contrast, in the US, the cost of military equipment is equivalent to 11% of manufacturing output (*ibid.*). More importantly, large-scale rearmament is a sensitive political issue in the country, not only because it may upset the military balance with the Soviet Union in the region, but also because of the Japanese's long-standing commitment to peace, banning of nuclear weapons and banning of exports of weapons. One has to remember that it was Japan who suffered Hiroshima and Nagasaki and lost the Second World War. Japanese perceptions of military forces therefore are different from those in the US which has hardly experienced war on its own territory this century. As a result, Japan must play a delicate balancing act and it is not surprising to see that, in spite of clear US interest in getting Japan to join SDI for instance, it is only recently, in the wake of the Toshiba affair, that the Japanese government has shown a more positive attitude toward Japanese participation in SDI, finalising an agreement which had been under discussion for months. In turn, the Japanese government expects that this step will help calm the tension created by the Toshiba sale of machine tools to the Soviet Union (*Financial Times*, 22 July 1987). In spite of all this, however, it seems unlikely that Japan will ever increase the relative weight of its military social force to anything like levels which will affect its present commercial performance in the international arena.

So, what does the future hold for the US social constituency of power? Leaving aside unpredictable events, nothing very different from what we are witnessing today. As the web of interests cannot be easily manipulated, it is most likely that contradictory trends will continue, only briefly masked by short term measures and conflicts with competitors. Resilience to change will almost certainly exclude timely adaptations among the social constituency, thus making a renewed crisis of confidence among the US people a very real prospect for the future. The time and the form of this crisis cannot be predicted but an essential ingredient is what some US analysts are even now predicting for the year 2000, namely, that 'The United States will eventually find itself subservient to Japan, seeking guidance and approval from the "rising sun" much as Western Europe now does from us' (Goldman, 1987, p.21). Clearly, such a dramatic change will be painful for the overthrown power and it is my view that as this occurs , the present complex of power will again be weakened while, simultaneously, the argument for the development of radical, humane alternatives will become stronger. But will it be enough for these alternatives to flourish, bringing a new humanely-oriented social complex of power? If we have to rely on past experience, the answer is no. There is no escape from the complexity of this problem but it is one that all those interested in improving society must face. At this point, we are back to some of the questions raised above. Where are the radical alternatives? Which are the social forces that will develop these humane alternatives? Most critically, which are the social forces that embody in their *raison d'être* the goals of humane ideals just as capital embodies the purpose of profit-driven accumulation?

Searching for a New Social Constituency

The concept of an alternative social constituency has preoccupied various scholars for many years. More than twenty years ago, for instance, Jacques Ellul raised this 'who-problem' when it was suggested to him that man has the potential to change society through the creation of large-scale social movements. 'Of whom are we speaking when we say "man"?' he then asked (Ellul, 1965). And in his answer, he could find no definite social force which would show a strong commitment to developing a different society. For him, the intellectuals, the technicians, the politicians, and workers are all involved in accepting the present form of development, while the individual is powerless and philosophers and artists do not count since they are non-technical (*ibid.*). Of course, Ellul was alluding to a process which he saw as even more fundamental than the control of technology for the purposes of capital or the military. In his view, this was the process whereby all social forces have been pervaded by the idea of technical progress, with the result that, whatever the influence of economic or other factors, technology has become an end in itself, developing according to its own internal laws (Ellul, 1967). This means that not only is the displacement of the present social complex of power necessary for a more humane alternative to develop, but, in the absence of any social force which has not been taken over by the concept of technical progress, a profound transformation in society's ideology is also necessary. In the words of Ellul (1965), 'What is needed is a kind of psychological and spiritual revolution affecting a considerable group of people' (Ellul, p.568). In 1972, Ellul was more specific, advocating Christianity as the answer: 'all must be turned to the glory of God. Nothing, either in the natural world or in the system created by man, can be left outside that glory.' (Ellul, p.236.)

Other scholars have differed with Ellul's theories, but the theme of 'conversion' has reappeared in different disguises in the attempt to provide a definitive answer to the 'who-problem'. Mumford openly stated that

> Nothing less than a profound re-orientation of our vaunted technological 'way of life' will save this planet from becoming a lifeless desert . . . For its effective salvation mankind will need to undergo something like a spontaneous religious conversion: one that will replace the mechanical world picture with an organic world picture, and give to the human personality, as the highest known manifestation of life, the precedence it now gives to its machines and computers (Mumford, 1970, p.413).

Denis Goulet (1977, 1983) expressed the matter in terms of the value system adopted by society. For him, if a particular society adopts, say, capitalism, then that society has also adopted a specific set of values which go hand in hand with capitalism. He points to infatuation with acquisition, mass consumption and endlessly wasteful changes as ingrained in the value system of a society where the search for 'the competitive edge' is a dominant factor. Furthermore, Goulet identifies in Western society the presence of a 'technological mentality', i.e. a set of values which are intrinsic to Western technology and which are implanted wherever this technology is adopted. Specifically, he distinguishes a particular approach to rationality and efficiency, a predilection for problem-solving and an exaggerated Promethean view of the Universe which leads adepts of technology to treat natural forces as well as human institutions as objects to be used and manipulated. This technological mentality pervades all social forces and reinforces capitalism, raising huge barriers to any radical change which might create an alternative ethical system in which humanity is given the central role. Like Ellul, Goulet sees the way out only in a general 'conversion' of society to a different set of values.

The economist John Kenneth Galbraith is not far from the same conclusion when he talks of the 'emancipation of belief' as a fundamental prerequisite to change the present form of development of society and technology. In *Economics and Public Purpose* (1974), Galbraith envisages the possibility of the state becoming a crucial instrument in society's effort to bring about a better social structure: one which fulfils the public purpose and not the interests of those currently in control of the power structure. For this to happen, the state must first be emancipated, i.e. be broken free from its role as instrument of those currently in control. Only then can its policies coincide with the public purpose. The state, however, can only be freed if the people themselves undergo the 'emancipation of belief', i.e. achieve a complete ideological break from the belief that the purposes and goals of the present system are indeed those of the individual. Searching for a social force which would embody power and achieve the public purpose, Galbraith finds the answer in the 'countervailing power of the consumer'. This is no different from saying in 'the power of the people' or any other categorisation encompassing those who do not belong to the dominant social complex of power but whose life is deeply affected by its control of society's development.

More recently, two of the most influential scholars in the field of technology in society at present, Langdon Winner and David Noble, have also touched on the 'who-problem' involved in the task of changing the present order and building an alternative, human-centred future. In his *Autonomous Technology* (1977), Winner identified a varied collection of social projects which would form a possible base for an alternative approach to future development. He did not mention any specific social force, but rather social movements such as the 'appropriate' or 'alternative' technology movement, the new arts and craft movement, urban and rural communes, alternative architecture, the peace movement, pioneers of radical software and new media, efforts to establish worker self-management in factories, etc. He recognised that as a common social approach, this collection of projects was still feeble and disjointed but it at least provided an indication that people had begun to question the roots of the current sociotechnical development shaping their lives. The enormous difficulties involved in bringing about a major transformation toward an alternative form of development were clearly spelled out by Winner; nevertheless, he placed great hope in the power of human reasoning and understanding as he proposed his 'epistemological Luddism' i.e. the practical dismantling of sociotechnical systems in operation as a method 'to learn what they are doing for or to mankind'. He added,

> If such knowledge were available, one could then employ it in the invention of radically different configuration of technics better suited to nonmanipulated, consciously, and prudently articulated ends (Winner, 1977, p.330).

Later, Winner's writing has shown a more pessimistic outlook. In *The Whale and the Reactor* (1984), for instance, there is no longer any mention of his 'epistemological Luddism' and the book ends on a tone of resignation about the present and a deep sense of uncertainty about the future. The following passage relates to a lecture on technology and the environment Winner gave in his home town. In this lecture, Winner argued against the placing of a nuclear reactor in Diablo Canyon, a beautiful spot near the town, and proposed that the community ought to take it over, dedicating it as a monument to the nuclear age. In its place a public park ought to be built. Winner then reflects on his proposal and the reaction of the people.

> Of course, my proposal was not taken seriously; too much money had already been spent, too much institutional momentum built up, too many careers invested, too many sermons preached from the pulpits of progress to allow any course of action that sensible. At present our society seems to prefer monuments of a different kind — monuments to gigantism, war, and the overstepping of natural and cultural boundaries. Such are the accomplishments we support with our dollars and our votes. How long will it be until we are ready for anything better? (Winner, pp.177-8).

Not surprisingly such a question is left unanswered, for it implies the need to reject the present outlook and be 'converted' to a new one which would make mankind the centre of societal and technical development. Before knowing how long this would take other questions need be considered; particularly, where within society is this process most likely to begin? and what can be done to stimulate its development? Winner does not pursue this line of enquiry. One reason may be because he does not use a framework of power relations identifying which are the dominant forces and, hence, who benefits most from the present situation.

Historian David Noble uses a power relations approach as he attempts to answer the two problems just raised. Noble contends that technology reflects power relations in society and hence its development is shaped by and for those interests in dominant control of society. Thus, the introduction of new technology by capital has reflected its own interests, often to the detriment of workers' interests. As the power of capital has grown, the task of challenging its control over society's and technological development has become more complex. Central to this situation is the role of the 'ideologies of progress' which have come to pervade society, and which effectively paralyse resistance to the present process in a web of promises about the future and arguments about efficiency, productivity and competitiveness. As a result, commitment to technological progress has also been adopted by intellectuals, trade unions and workers, despite the fact that a capital-controlled technological progress normally means a great deal for the interests of capital and other dominant forces and very little for anybody else. It is among workers, however, that Noble sees a possible source of rebellion. Indeed, writing about direct action in 1983, he spelled out his belief that the resurgence of rank and file action at the time was pointing to something fundamental. 'Having overcome the fatalism of technological determinism', he wrote, 'they have now begun to overcome also the futurism of technological progress, and to shift attention back to the present' (Noble, 1985, p.148). Noble then called on responsible intellectuals to follow the lead given by workers who had begun to smash the physical machinery of domination. Intellectuals 'must begin deliberately to smash the mental machinery of domination' (*ibid.*). They must help break the 'hypnotism' of the ideologies of progress which had led society astray from human-oriented development. Of course, in pursuing this end, confrontation with power and domination is inevitable since these ideologies of progress, like technology itself, reflect the interests of those in control.

Indeed, having examined the shape and direction of the current drive for automation in the US, Noble (1984) advocates that 'resistance to the current technological assault' and 'direct political confrontation against those now in power' are steps in the right direction. The same is valid for all other efforts to develop alternative, human-centred forms of technology. Ultimately, however, Noble agrees with Mumford in that, in themselves, these steps would not suffice; that a transformation of the ideological basis of the whole system is necessary for a genuine and lasting transformation of societal and technological development. This is the real challenge involved in the radical alteration of the path of the present technological revolution.

> Clearly this is an extraordinary challenge which would require, among other things, a fundamental rethinking of the form and function of science and technology as well as the formulation of a practical vision of a more democratic, egalitarian, humane, creative and enjoyable society (Noble, 1984, p.353).

Above, we indicated that, during 1983, Noble had seen signs that the rank and file in the US had begun to overcome the grip of the ideologies of technological progress. Today, with hindsight, it is clear that such a development has not spread, and has failed to make any impact. Whether and when it will grow nobody can tell, for we are again talking here of the general 'conversion' of society to a radically different outlook. Certainly, one could always take one or other partial development and make optimistic forecasts for the future. In the present conditions, however, given the resilience of the complex of power and its deep-rooted ideological foundations, it seems hardly possible to embark on hopeful predictions. In this respect, Noble's closing words in his *Forces of Production* (1984) are more realistic. We see, he states, 'not the hopeful hymns of progress but the somber sounds of despair and disquiet'. (Noble, p.353).

These are the realities, issues and challenges involved in achieving a human-centred development of society and its technology. These are the realities facing the future of the microelectronics revolution. All the distinguished scholars we have been looking at agree in their conclusion that, whatever the form of the changes aimed at, the fundamental prerequisite is a deep and generalised ideological transformation which emancipates society from the grip of the ideologies of progress and 'converts' it to a world outlook that puts humanity (God in Ellul's view) at the centre of development. All authors also coincide in acknowledging the enormous difficulties implicit in this transformation; but all reaffirm that, although difficult, this transformation is not impossible. Thus there are hopes for the future. After all, history has witnessed many strong power complexes rise and fall as well as great ideological transformations occur.

> This order of change is as hard for most people to conceive as was the change from the classic power complex of Imperial Rome to that of Christianity, or, later, from supernatural medieval Christianity to the machine-modelled ideology of the seventeenth century. But such changes have repeatedly occurred all through history; and under catastrophic pressure they may occur again (Mumford, 1970, p.413).

It is not my intention to refute here what these scholars have agreed upon. Thus, as we return the focus of our analysis to the present form of development of microtechnology and the R&D system in the US, we know that the current complex of interests making up their social constituency, although strong, is not impregnable, and that the ideologies of progress reinforcing its control, although pervasive, are not necessarily permament. Both may eventually collapse, thus opening the way for a new social

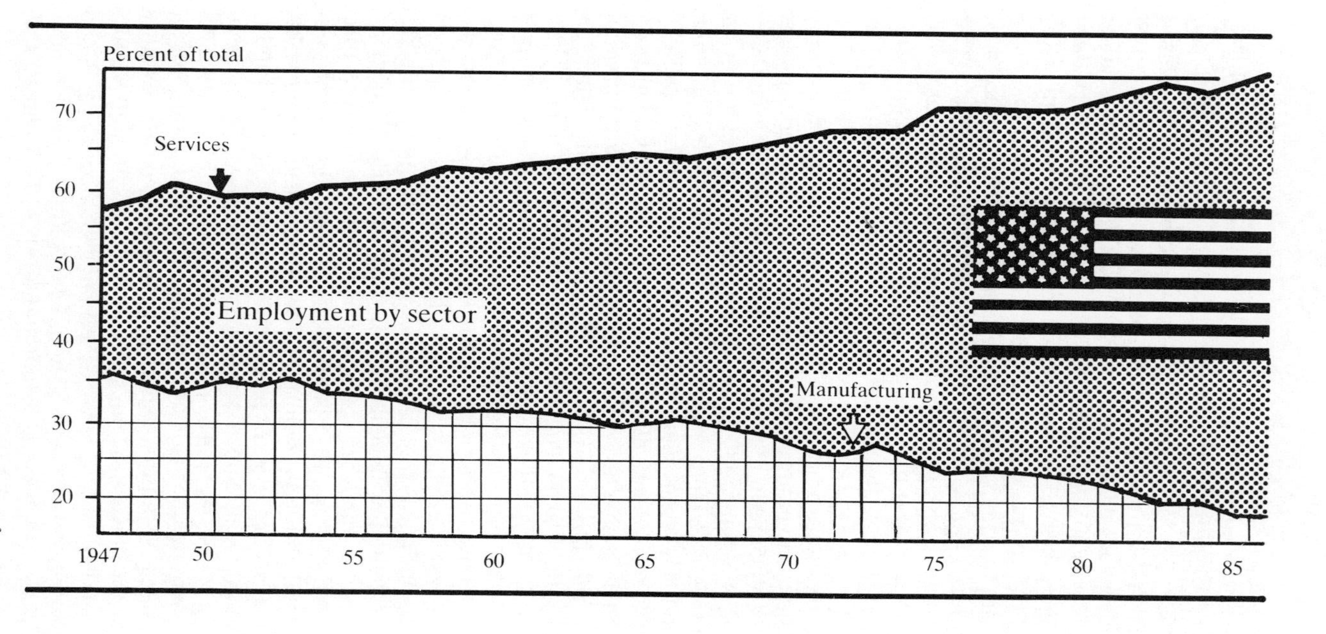

Figure 41. Evolution of Employment in the United States, by Economic Sector (1947-85)

Source. Financial Times, 22 May 1987, p.16

constituency and new, more humane goals. At present, however, when the existing complex of power still looks unassailable and the ideologies of progress penetrate all society, it is difficult to envisage which social forces will lead a new social constituency or even the struggle against domination of the present one. Langdon Winner has identified a collection of critical social movements as a possible reservoir for a new alternative development. David Noble, in a view which follows the Marxist tradition of social revolution, seems to put his faith in workers and intellectuals. The workers because they, more than anybody, suffer the direct consequences of societal and technological development ruled by profit, military power and control; the intellectuals because they can help break the hold of the ideologies of progress by revealing their human irrationality and role in entrenching the present complex of domination.

In time, it may well be that the forces identified by both Winner and Noble will combine to create the basis for a new alternative. This is impossible to predict, however, since the societal process is fluid, constantly changing the content and relative strength of its social forces. One development to watch closely in this respect is the transformation of the working class as a result of the increasing automation of the manufacturing base of society. The trend, as illustrated in Figure 41, is for a relative decline of the proletariat engaged in manufacturing, the class Marx saw as the most likely to revolutionise a capitalist social order which would eventually fetter the development of productive forces. The problem now is that, with microtechnology, the development of productive forces under capitalism has begun to undermine the strength of the proletariat in a process which looks like leading to a decline similar to that caused by mechanisation among the population engaged in agriculture.[1] Indeed, the result of this process will depend on the struggles which are being waged at the moment. In this respect, it is plain that although one may speculate about the future, today it is among the social forces identified by Winner and Noble that the seeds of hope for a better future lie. It is their resistance to the designs of the power complex, their search for new forms of technology and way of life which are creating a valuable stock of experience which, however piecemeal today, may eventually become the platform of a more humane tomorrow. I believe with Mumford, however, that it will be through catastrophe that society will develop a new vision of itself and its future. As Mumford put it,

> something like a universal awakening sufficient to produce an internal readiness for a profounder transformation must

> take place. Such a reaction, one must honestly confess, has never yet occurred in history solely as a result of rational thinking and educational indoctrination: nor is it likely to occur in this way now — at least within the narrow time limits one must allow, if greater breakdowns and demoralizations are to be circumvented . . . something like catastrophe has become the condition for an effective education (Mumford, 1970, p.441).

Mumford was optimistic that this is not 'a dismal and hopeless conclusion' because 'the power system, through its own overwhelming achievements, has proved expert in creating breakdowns and catastrophes' (*ibid.*). He believed that in its lack of humane concern, and its concentration on power, profits, productivity, prestige, the power complex has drawn humanity towards catastrophe as the accumulation of conflicts and irrational exploitation of the planet proceed unchecked. The words 'ozone layer', 'acid rain', 'greenhouse effect' come quickly to mind here. Consequently, Mumford saw the possibility of a new, human-centred beginning for 'If mankind is to escape its programmed self-extinction the God who saves us will not descend from the machine: he will rise up again in the human soul' (*ibid.*, p.413). On the other hand, we also know that it was the catastrophe of the Second World War that re-invigorated the present social complex.

The kind of catastrophic situation envisaged by Mumford as necessary for a wholesale 'conversion' of people to a new way of life is at present an undreamt-of possibility. People tend to grow used to visible signs of irrationality and enormous human suffering, ignoring the fact that the means to remedy many social maladies are already available.

We have said that the US social complex of power is navigating difficult waters and heading for a possible crisis of confidence in the future. But this is highly unlikely to reach the proportions referred to by Mumford, whose vision is one of planetary crisis. Most likely, cushioned and veiled by the ideologies of progress, this crisis will lead to an internal realignment of the present social constituency of power, which may re-emerge as the guarantor of technological progress. In this vein, one may reason that only if the social complex of power could no longer guarantee technological progress, would its stability be effectively undermined. Although in itself this would not ensure the emergence of democratic and human-centred forms of societal development, at least the present order would have shown itself unable to fulfil the technological goals that justified its existence. Was this not what Marx was thinking when he wrote that ' no social order is ever destroyed before all the productive forces for which it is sufficient have been developed' (Marx, 1975, p.426)?

APPENDIX

RISE AND DEVELOPMENT OF THE US'S R&D SYSTEM PRIOR TO THE SECOND WORLD WAR

By the late nineteenth century, the concentration and centralisation of capital, which Marx described in *Capital*, had taken capitalism into its monopoly stage[1] and the related state of imperialism [Lenin (1944), Sweezy (1970)].[2] That is to say, within the most advanced capitalist nations the rise of monopolies, which Baran and Sweezy (1975) date from about 1870, had signalled the end of the predominantly competitive stage of capitalism.

In the new stage, the dominant economic unit was to be the corporation, a powerful instrument of centralisation and accumulation of capital[3] which would eventually grow into the present-day giant corporation commanding large economies of scale and exercising oligopolistic control of extensive markets.[4] In Baran and Sweezy's words, under monopoly capital, the 'dominant element, the prime mover, is Big Business organized in giant corporations' (Baran and Sweezy, 1975, p.62).[5]

This brief introduction outlines the industrial-economic set-up which saw the rise of the first science-based technologies and industries, namely, the electrical and chemical industries based respectively on scientific knowledge of electromagnetism and chemistry.[6] As we shall see, from the start the monopolistic advantages associated with the possession of inventions protected by patents found a most powerful resource in the application of scientific knowledge which became a means of constantly ensuring and extending such monopolistic advantages. In the science-based industries, therefore, a mutually beneficial relationship developed whereby scientific knowledge was pressed into the service of corporate capital providing it with a constant source of productive growth and expansion, while corporate capital became the social framework shaping and creating the specific form of integration of the basic social constituents of science-based technological and industrial activity. As Noble put it in relation to the US,

> the history of modern technology in America is of a piece with that of the rise of corporate capitalism. Both contributed to a transformation of the *modus operandi* of industrial capitalism — the one providing the wherewithal for unlimited productive growth by implicating science in the production process, the other offsetting the destructive tendencies in an unchecked competitive market economy by making possible the regulation of production, distribution and prices (Noble, 1977, pp.xii-xiii).

The Emergence of the R&D System

Along with the incorporation of science into the sphere of technological and industrial activity there emerged a set of institutions whose purpose was to enable such a process and which has become known as the research and development (R&D) system. The origins of this system may be traced back to the eighteenth century, to the formation of the first technical schools in France and Germany[7] and, more directly, to the introduction of the experimental technical laboratory at the *Technische Hochschule* in Munich in 1868.[8] With the emergence of the laboratory, technology finally became a field of research in itself (Weingart, 1978), incorporating science to supply the needs of the emergent science-based industrialisation.

It was in Germany, during the last decades of the nineteenth century, that the incorporation of science into industry was most developed. There, a policy of fostering science and its industrial application was to develop in a way unknown to other contemporary industrial powers, notably Britain which remained in the grip of empiricism (Bohme *et al.*, 1978). As Braverman has described,

> at a time when British and American industry used university-trained scientists only sporadically, for help on specific problems, the German capitalist class had already created the total and integrated effort which organized, in the universities, industrial laboratories, professional societies and trade associations, and in government-sponsored research a continuous scientific-technological effort as the new basis for modern industry (Braverman, 1974, p.162).

The clearest demonstration of the above situation was the rise of

Germany as undisputed world leader in the chemical science-based industry by the turn of the century. In effect, although, as Landes (1969) has argued, the first years of the new branch of chemical manufacturing saw Britain first, with France in second place,[9] by the late 1870s German industry had captured about half the world market and, by the turn of the century, its share was around ninety percent.[10] The extent to which such a monopolistic supremacy depended considerably upon the systematic integration of science into the industrial process is confirmed by Freeman.

> The German industry in the 1870s had already established the new pattern of in-house R&D leading to the introduction of new products and processes. Bayer, Hoechst and BASF (Badische Anilin und Soda Fabrik) were among the first firms in the world to organize their own professional R&D laboratories (Freeman, 1974, p.48).[11]

The experience of the chemical industry, and also that of the other science-based industry of the nineteenth century, the electrical industry, had momentous significance. It firmly established the importance of science for industry and capital accumulation. Indeed, in Hobsbawm's view,

> by the end of the nineteenth century it was already clear . . . that the output of technological progress was a function of the input of scientifically qualified manpower, equipment and money into systematic research projects (Hobsbawm, 1978, p.174).

Put in other terms, by the end of the nineteenth century, the integration of human, material and financial resources intrinsic to all technological development began to revolve, as never before, around the systematic production and/or application of scientific knowledge through the R&D system.[12] More importantly, as this process was mediated by corporate capitalism, it is true that, as Noble (1977) argued, from its inception the development of the R&D system and, more generally, of science-related technologies came to play a major part in the process of capital accumulation and monopoly power. In turn, the latter process also helped to shape the development of the R&D system itself.

The economic sphere was only the most immediate place where the importance of the R&D system manifested itself. In time, the system's role was to reach deeply into other areas of society, thus bringing other powerful social interests (e.g., military, government) to wield influence on its development. In this context, as shall be seen below, the control of basic human, financial and material resources of the R&D system will expand beyond industry in such a way that it will be the convergence, conflicts and relative weights of all the social interests involved that will explain its development. Undoubtedly, industrial capital will still shape the development of the R&D system, but now in conjunction with other social interests with whom it will share control of the R&D system's basic resources.

By focusing on the particular case of the US, I shall look at the social forces, and their interrelations, which have been involved in the origins and development of the R&D system and hence, of science-based technology. In so doing, I shall illustrate my discussion with particular evidence from the science-based industries most directly related to today's microtechnology. Thus, I expect to gain an insight into the historical roots of the social constituency of United States' microtechnology. I shall start by focusing my attention on those forces and tendencies coming from within the industrial and productive sphere, for it was here that the R&D system and science-related technologies began their development in earnest.

The Emergence and Development of the US's R&D System: Corporate Capital as its Dominant Social Constituent

In the US, where the industrial R&D system has achieved a leading position in the twentieth century, the first research laboratory set up for the specific purpose of systematic invention was organised by T.A. Edison at Menlo Park, New Jersey, in 1876. According to Lindsay,

> This was the forerunner of the modern industrial research laboratory, which has revolutionized the relations between science and technology in the twentieth century . . . [however] . . . Edison's laboratory was not a laboratory for scientific research. His sole purpose was to dream up and then produce gadgets which would have economic value, i.e., could be sold at a profit to a public which found them useful or exciting . . . Nevertheless, his laboratory had a well-defined program and pursued it systematically (Lindsay, 1973, p.216).[13]

Towards the end of the nineteenth century and beginning of the twentieth century the number of commercial research laboratories began to increase as the importance of R&D began gradually to be recognised. In 1886, Arthur D. Little, an applied scientist, started his independent firm. Others followed: Eastman Kodak (1893), B.F. Goodrich (1895), General Electric (1900), and Du Pont (1902) were some of the earliest manufacturing firms to establish laboratories; and the Bell Telephone System (1907) was among the first utilities to do so (Mansfield, 1969). Also, the first government laboratories were established in 1887 by the Department of Agriculture (Braverman, 1974).

Most scholars agree that before the twentieth century, R&D activity in industry put little emphasis on fundamental research. Indeed, it is a matter of agreement that organised industrial research, involving fundamental research, began in 1900 at General Electric's (GE) laboratory at Schenectady, New York, the result of the company's conscious policy to make scientific knowledge a major part in its development [Barlett (1941), Lewis (1967), Allison (1980)]. In Lewis's words,

> Here under the leadership of Willis R. Whitney, formerly chemistry professor at MIT, a large staff of scientists and technical personnel was formed and kept constantly at work on a variety of problems involving electrical phenomena. By engaging in continuous education, the company helped to keep abreast of any new developments in its area of commercial interests, thus maintaining its dominance and guarding against future competition (Lewis, 1967, p.626).

The laboratories organised by Du Pont in 1902 and Bell in 1907 followed GE's path, as did other industrial laboratories set up later.[14] The reasons behind such change were both technical and socio-economic but an immediate factor was fear that the scientific knowledge which had underlain the rise of the science-based industries in the nineteenth century was no longer sufficient for the demands of further technological progress.[15] Thus, the frontiers of scientific knowledge had to be pushed forward and, as it was apparent that they could be done so indefinitely with obvious implications for the future of the companies, a policy of in-house R&D activity, including fundamental research, was seen by corporate capital as an important asset in the long-term commercial strategy of science-based industries.[16]

The industrial laboratories that emerged from a policy of in-house R&D, however, demanded considerable financial resources and, in practice, could be afforded by only a handful of powerful industrial concerns, namely, the large corporations. Such a development marked the end of the early period of the industrial laboratory dominated by the figure of the inventor-entrepreneur. He had clearly established the foundations of science-based activity,[17] but by the early twentieth century, as Birr (1966) put it, it 'was left to the large corporations developing everywhere in the American scene . . . to introduce organized industrial research' (Birr, p.68). The reason was simple: it was these large corporations where there was 'sufficient financial resources and stability to support the laboratories and where there was a rapidly changing, competitive technology which made successful research imperative for the sponsoring firm' (*ibid.*).[18]

In this way, the early twentieth century saw large corporate capital become the dominant social constituent behind the development of the US's R&D system. This meant that the system's shape and direction became largely influenced by the needs of the particular interests of corporate capital. In this respect, profits, competition and the search for monopoly through patenting and innovation became prominent factors behind the development of the science-based technology. In trying to systematise the process whereby all these factors came together, mostly within the industrial sphere of society, the following major developments can be distinguished.

a) Incorporating science into the industrial sphere of society was pioneered by those industries which from the outset had depended upon science for their development. These were the chemical and electrical industries which, as seen earlier, were the first to incorporate R&D within their business. At the beginning, the industrial laboratory was used chiefly for development and some applied research, but large-scale organised industrial R&D, involving fundamental research, began in the science-related industries early in the twentieth century.[19] This example would later extend to other industries. In Barber's words, 'Older industries were much less quick to bring science into their activities' (Barber, 1970, p.214).[20]

b) From the start, in-house R&D activity was an integral part of the process of capital accumulation in science-based industries. As a consequence, this R&D activity was intimately bound up not only with the specific technical problems facing each industry but, indeed, with the very essence of the process of capital accumulation, namely, profit-making. In the market context, this meant that competition, or rather the constant search for monopolistic positions, came to play a major role in shaping and stimulating specific technical challenges and hence the specific content of industrial R&D. Birr has explained how this process took place in the R&D context.

> The late nineteenth century was a period of bitter competition, deflation, and business upheaval in America as industrial leaders desperately searched for security and stability. The search most obviously led to methods of lessening competition, including interfirm cooperation, price fixing, mergers and other activities which so aroused the ire of the antimonopolists. But in some cases the search for corporate security led towards science and technology . . . Such needs were most deeply felt in those industries characterized by rapidly changing technologies and particularly in those industries whose technologies were dependent on science. It was no accident that the modern industrial research laboratory first emerged in industries such as communications, electrical machinery and chemicals. The first two had never had a craft tradition in advance of scientific knowledge; indeed, electricity had never been of practical use before scientific understanding of the principles on which it operated (Birr, 1979, p.197).[21]

The major role played by social factors should not belittle the importance of the specific technical demands essential to the development of science-based industries. For, while it is true that technical progress has no impetus of its own, it is equally true that competition, or more generally, the accumulation of capital, cannot completely explain the specific form and needs of the R&D system at any particular time. It seems clear that the R&D system also depends heavily upon both the current state of technology and the scientific-technical requirements inherent in the firms' overall business strategy. Hoddeson, for instance, gives technical factors an important role in explaining the incorporation of scientific research in the R&D activity of the Bell System during the early twentieth century. She suggests the following three principles:

> (1) nonscientific objectives lead the company to particular technological problems; (2) technological problems so profound or complex that the usual approaches to such problems fail, leading the company to seek deeper understanding of the underlying physical phenomena; (3) research that is successful in technological terms reinforces the company's commitment

> to scientific studies in the particular area (Hoddeson, 1981, p.516).[22]

c) By pushing forward the frontiers of scientific-technical knowledge, industrial R&D activity brought about, more than ever before, the likelihood of two developments which might threaten the market position of established firms.

(1) loss of technological control and competitiveness to other firms in the field, eventually leading to loss of market and fall in the rate of profit.

(2) complete or partial technological and market displacement as a result of radical innovations bringing about new and more competitive products and/or processes.

The effort to prevent both possibilities from materialising became a major factor in shaping the technological policy and, particularly, the R&D policy of established firms. To begin with, pioneering science-based concerns sought to protect their technological leadership by using the monopolistic protection provided by the patent system.[23] Here they would try to develop, or simply, acquire the rights to all those patents relevant to the technological control of the business. Indeed, as Noble has described for the case of GE and AT&T,

> [their] . . . individual policies . . . were carefully designed to gain and prolong monopolies over patents vital to their industry. Toward the end, they employed such methods as incomplete disclosure of information in patent applications, the use of trademarks, the outright suppression or delayed introduction of patented apparatus, the compulsory assignment of employee patents to the company, and the deliberate production of auxiliary patents (Noble, 1977, p.93).[24]

In this way, big corporations could monopolise the market and, at least for the duration of the patent, very little could take place which was technically out of their reach unless some major innovations were to alter the technical scene. In practice, however, this policy could not offer long-term security to companies since not only did patents last for a limited period[25] but the possibility of competing technical developments was quite real and could not be effectively prevented by a policy chiefly emphasising the legal aspects of patents while overlooking their systematic production through organised R&D. Corporate capital realised that without this production, it might in time face loss of technological control or leadership of the industry or, simply, technological displacement as a result of development in other fields. The only effective way of insuring against such possible threats was to maintain a constant policy of industrial R&D seeking to keep abreast with most, if not all, aspects of technological progress relevant to the industry.[26] This became particularly true as the technical challenge from the commercial strategy of the firms grew more complex. For, as seen, the more complex the technical challenge, the greater the need for scientific research and, ultimately, for a well-organised R&D system. This was the stage that large science-based corporations had reached in the US by the early twentieth century, and which led them to organise the first industrial laboratories incorporating science as a permanent activity [Noble (1977), Hoddeson (1981), Birr (1979)].

d) The success of the science-based corporations and the growing R&D effort brought with it two important developments which not only expanded the industrial R&D system beyond large corporations into smaller businesses, but also brought universities within its sphere of influence.

The first development was the diffusion of the importance of R&D for market success. The second development was the growing costs of such R&D activities as the range of scientific disciplines involved in the technological process grew. As we saw earlier, in-house R&D laboratories were expensive, and only large corporations could generally afford them. With increasing technical complexity and costs this problem worsened since 'even in those great enterprises which were able to support some fundamental as well as applied research (notably Du Pont, GE, and the Bell System), in-house industrial laboratories were not able to meet all their research needs' (Noble, 1977, p.121). In addition, there had always been one vital requirement that no industrial laboratory could satisfy, namely scientifically trained manpower who could maintain and expand the R&D system.[27] Because of this, wider social resources had to be tapped to ensure the further development of the R&D system and satisfy the increasing needs of industry. This led to closer relations between industry and the university, and to the emergence of several outside research institutions. In effect, as Dickson has noted in relation to universities,

> The industrial interests in science also meant drawing universities even more closely into the corporate sphere of influence, since they were soon recognized as offering the most likely source of ideas required for innovative products and processes (Dickson, 1984, p.62).[28]

Outside research organisations took various forms which have been well documented elsewhere.[29] The most important were university research facilities, trade associations laboratories, non-profit research institutes such as the Mellon Institute, commercial research laboratories such as the A.D. Little Laboratory, private foundations such as the Carnegie Institution, federal government research facilities in government bureaus such as the National Bureau of Standards and also, particularly after the First World War, military department research facilities. All these institutions, together with the facilities of colleges and universities and in-house R&D laboratories, greatly expanded the institutional network of human, financial and material resources at the service of industrial R&D and ultimately, of course, at the service of corporate capital exercising hegemonic control of such a process.

Within a few decades, therefore, industrial R&D had come a long way from the pioneering efforts of inventor-entrepreneurs, attaining an institutionalised system involving wider social interests than just industry. Yet, before the First World War the development of the US's R&D system had revolved mainly around industry dominated by corporate capital. The impact outside the industrial sphere had been limited, reflected in the small relative influence of other social interests in the social constituency shaping the development of the R&D system. With the advent of the First World War, however, all this began to change as the pressures of war involved powerful new social interests in the development of science-based technology. The main result of this was the consolidation of a complex of social interests in which capital was only one of its constituents alongside the interests of government, the military and science. This new social constituency created new problems, with the pressures of war binding the diverse social constituents, also inevitably bringing conflict. And at the same time, a new dynamic for the R&D system also emerged.

The First World War and the Widening of the Social Constituency of the US's R&D System

Until the end of the First World War, the process which we have been examining above proceeded rather slowly. Thus, by 1920 the number of companies carrying out R&D activities was reported to be around 300. As Barlett commented, 'a small figure when compared with the number of companies for which research was a sound undertaking' (Barlett, 1941, p.34). Moreover, there was a concentration of industrial R&D activity in the science-based industries, with two-thirds of all research workers employed in the electrical, chemical and rubber industries (*ibid.*).

With the outbreak of the war the strategic importance of science and technology became apparent (Schroeder-Gudehus, 1977). The crucial role of the science-based technologies in military power and national defense was firmly demonstrated by, for instance, the widespread use of chemical weapons in the battlefield[30] and by the shortages, particularly of chemical products and equipment following the cut in exports from Germany. In the US, this led to greater government and military involvement in the R&D system which so far had been developing largely under the unchallenged control of corporate capital.[31] Although in the past government and the military had been involved in the development of science and technology,[32] such activities were small-scale and did not represent any consistent interrelation with science and industry. According to Lasby, for instance, early relationships between science and the military were superficial, marked by antagonism and tension, so much so that, by the turn of the century, 'the separation of science from the services was nearly complete' (Lasby, 1966, p.258). In addition, it was also the case that in the US 'the military utilization of science to improve or develop weapons was virtually nonexistent until World War I' (*ibid.*, p.252).

The war, however, provided the impetus and context within which the first links of a systematic relationship between science, industry, government and the military were forged. This was the beginning of the broadly-based complex of social interests which, by controlling the basic resources of the R&D system, has come effectively to shape the development of US science and technology. The main practical effects of the war upon the development of the R&D system have been described by Pursell.

> World War I left its mark on American science. Research, though not a complete stranger, was finally established as a strong partner of both industry and the military, and its position in the government generally was strengthened. Furthermore, the kind of cooperative assault on large problems which had been the developing technique of government bureaus for decades, now became a common experience for many American scientists. And finally, the war had left behind a whole string of new institutions, from the National Research Council [formed in 1916] to the nearly aborted Naval Research Laboratory [finally set up in 1923], which would have a vitalizing effect on government science in general (Pursell, 1966, p.237).[33]

The war clearly expanded the social constituency of the US

R&D system beyond the realm of corporate capital interests. More crucially, the war saw a considerable alteration in the relative power wielded by the social constituents. In particular, corporate capital lost much of its influence as the impetus of war supported the relative position of military/government interests. These came effectively to shape the development of the US's R&D system during those years. The greatest effect of such change was immediately felt as military objectives swiftly displaced purely economic goals as the dominant preoccupation of the country's R&D pursuits. This situation was clearly depicted by Barlett when he wrote that, by the time of the armistice, 'practically every scientist possessed of any capacities for research had been called upon to aid the country with his special knowledge' (Barlett, 1941, p.35). To the changes within the social constituency of the US's R&D system, therefore, there had been added changes in the system's social role and, consequently, in the final product of its activities.

These developments highlighted some important aspects of the R&D process. On the one hand, they revealed the profound social nature of the R&D system's development and hence, the fallacy of the idea that technology possesses a 'life' of its own. On the other hand, they showed the contradictions in this process insofar as changes in its social constituency, by channelling development into particular directions, are bound to conflict with its progress in others. During the war, for instance, as Lewis describes,

> The drain of scientific talent into projects of a military or semi-military nature . . . caused the temporary cessation of research in various areas of peacetime application and interfered with the progress of fundamental exploration in basic science (Lewis, 1967, p.628).[34]

This means that, while diverse social interests could converge in their support of the R&D system on the common basis of wartime military concerns, the military use of the basic resources of the R&D system tended to hinder progress in areas of mainly civilian application.[35] After the war, however, this military emphasis lessened as the waning force of war was accompanied by a major decline in the military's influence on wartime R&D activities.[36] As this took place, however, a new basis for the peace-time convergence of social interests emerged, namely, international economic competition.

The war had greatly improved the position of US industry in the world market.[37] This meant that, in its aftermath, international economic competition became a major preoccupation and hence, a major new factor capable of galvanising a social constituency behind the development of the US's R&D system.[38] This postwar social constituency, however, could not be the same as the wartime constituency which had been dominated by military interests. Indeed, I have already said that the relative weight of military interests declined significantly after the war. In the absence of military pressures, the wartime social constituency had become largely artificial and this was expressed in changes in the degrees of influence held by the various social interests upon the development of the US's R&D system.

Under new postwar conditions, the dominant social constituency of the US's R&D system reorganised itself on the basis of international economic competition. This meant that a new convergence of social interests emerged, bringing together mainly the interests of capital, science and government. The arguments of prominent scientists of the wartime social constituency of the R&D system leave little doubt as to where the foundations of the new postwar social constituency were laid.

> George Ellery Hale, the astrophysicist, Robert A. Millikan, the physicist, and other influential scientists used the argument that the success of research in industrial products depended on the acquisition of new knowledge attained through pure scientific research. American industry, they claimed, might be endangered by postwar competition unless aid were given to the basic science on which new products depended (Coben, 1979, p.232).[39]

The pressures of international economic competition, however, could not offer the same powerful effect of wartime military pressures. This can be seen, for instance, in the reduction in size and influence of the NRC as a result of the decline in military and to a lesser extent government involvement which followed the end of the conflict (Swain, 1967).[40] Thus it was left primarily to industry, where the pressures of competition and search for monopolistic positions were an immediate concern, to retake its prewar role as the dominant social constituent of the R&D system. The context in which this took place, however, scarcely resembled the prewar situation, as war had broadened the base and scope of development of R&D in the US.[41] In particular, there was greater appreciation of the value of R&D. For the first time, a national research coordinating organisation, the NRC, was created. After the war, although reduced in scale, the NRC turned its attention to peace-time problems, providing an institution not just for supporting research but for the coordination of the various institutions involved in the

R&D system (Noble, 1977). In this context, according to Lindsay (1973), after 'the mid-1920s the claim of basic research to be an integral part of the program of every industrial research laboratory was uncontested'. At the same time, there was a sizeable increase in the number of industrial laboratories and personnel engaged in R&D activities. From 300 industrial research laboratories in 1920, the number rose to more than 1,000 in 1927, over 1,600 in 1931, and 2,200 in 1940 [Coben (1979), Birr (1979), Barlett (1941)]. Correspondingly, the number of personnel employed in industrial laboratories also increased markedly: from 9,300 in 1920 to 33,000 in 1931 and 70,000 in 1940 [Birr (1979), Barlett (1941)].[42] Finally, the number of large-scale research organisations also increased. 'Only 15 companies had research staffs of more than 50 persons in 1921; in 1939, there were 120 such companies' (Barber, 1970, p.214).

These increases did not mean any great change in the development pattern of the US's industrial R&D system. Thus, by 1940, three-fourths of all its professionals were trained as chemists and engineers, emphasising the dominance of science-based industries [Cooper (1941), Barber (1970)]. On the other hand, although industrial research had spread to companies of all sizes, by 1940 most of the effort was still conducted by a 'rather limited number of large corporations' (Cooper, 1941, p.182). This fact highlighted the continuing hegemonic control of the human, financial and material resources of the industrial R&D system by large corporate capital. In other words, while capital in general became the dominant social force within the post-First World War social constituency of the US's R&D system, it was in fact large corporate capital, particularly of the science-based industries, which exercised the commanding role in the system's development. In general, this situation remained until the outbreak of the Second World War when military pressures once again changed conditions, reviving the wartime social constituency of power and consequently initiating a new period in the history of the US's R&D system.

Notes to Chapter 1

1. In this respect, one of the most sensitive issues relates to the problems facing the scientific community arising from the apparent contradictions between freedom to pursue research for common good and the very concrete interests and demands of capital and the military for instance. Some authors have touched upon this issue [David (1979), Stanfield (1980), Prager and Ommen (1980), Lasby (1966)], but here we shall assume Noble's view made explicit in the following passage. 'First and foremost, the very fact that scientists and engineers are in a position to learn about the properties of matter and energy and to use their knowledge for practical ends, to make decisions about the design and even the use of the new technologies, indicates their close relationship to social power. Such ties to power afford them access to the social resources that make their achievements possible: capital, time, materials and people. Thus, it is no accident that technical people are often allied so closely with the owners of capital and agencies of government; the connection is the necessary prerequisite of scientific and technological development, given the social relations of American capitalism' (Noble, 1984, p.43).

2. Adams (1982) has studied this interpenetration of the military, government and large corporate capital where it has become more conspicuous, namely, in the defense industry. He has concluded that 'Over the years the defense industry has become a *de facto* participant in the policy-making process. As in other areas dominated by powerful corporate interests, a policy sub-government or "iron triangle" has emerged . . . Political scientists describe an "iron triangle" as a political relationship that brings together three key participants in a clearly delineated area of policy-making. The Federal bureaucracy, the key committees and members of Congress, and the private interests. In defense, the participants are the Defense Department (plus NASA and the nuclear weapons branch of the Department of Energy); the House and Senate Armed Services Committees and Defense Appropriations Subcommittee, as well as Congressional members from defense-related districts and states; and the firms, labs, research institutes, trade associations, trade unions in the industry itself . . . The special interests and the Federal bureaucracy interpenetrate each other. Policy-makers and administrators move freely between the two areas and policy issues are discussed and resolved among participants who share common values, interests, and perceptions.' (Adams, 1982, p.24). See also Dickson (1984).

3. The practical way in which these interests interlock around R&D for military purposes in the US is described by Sapolsky (1972). 'In the United States, though there are government-managed weapons laboratories and arsenals, the private sector is a significant participant in the weapons acquisition process. Universities are actively involved in the conduct of military research. Non-profit corporations advise on design and management questions. And business firms perform research, development, test, evaluation and management functions as well as manufacture weapons.'

(Sapolsky, p.453). See also Norman (1981).

4. In this context, science is used in its more common definition, namely, the pursuit of systematic knowledge of natural phenomena (Mulkay, 1977). According to Freeman (1974), 'the expression "*science-related*" technology is usually preferable to the expression "*science-based*" technology with its implication of an over-simplified one-way movement of ideas' (Freeman, p.29). On the other hand, the expression "*science-based*" reflects the fact that scientific knowledge is indeed fundamental to the very existence of the technology.

5. This is not to suggest that everything will be exactly the same for different science-related industries. In fact, every industry will exhibit its own particular characteristics in accordance with its importance vis-a-vis the interests of the diverse social constituents. The aircraft and missiles industry, for instance, has consistently had paramount military importance and hence, it has been heavily influenced by the role of the military. See Kaufman (1972). This industry has consistently received the highest proportion of Federal funds for industrial R&D and such funds constitute a very high proportion of that industry's total R&D expenditures. At its highest during the early 1960s Federal support reached ninety-one percent of the total (NSB, 1981).

6. In practice, the distinction between 'basic' and 'applied research' is not a clearcut one and various authors have pointed to the lack of sharp boundaries between the two [Mulkay (1977), Mansfield (1972), Freeman (1977)]. Similarly, there is an awareness of the serious limitations of the linear hierarchical model portraying technological innovation as leading from basic research to applied research and development [Layton (1977), Barnes and Edge (1982)]. It is more generally accepted that the relationship is in fact one of a dialectical interaction where the search for the solution of technical problems leads to problems in basic research and vice-versa. Furthermore, in a more general context, there is evidence that by and large, the search for science-related technological developments in particular directions is reflected in the selection of areas of pursuing basic research. Thus, although basic research is still about understanding natural phenomena, its connection with applied R&D is a very real one from the point of generating technical innovations. Freeman (1977) suggests that this kind of research may be called 'background' or 'oriented' basic research.

7. For other authors dealing with the concept of R&D, see Freeman (1977), Hoddeson (1981), Pavitt and Worboys (1977).

Notes to Chapter 2

1. The Appendix provides a historical analysis of the origins and pre-Second World War development of the R&D system, looking at the variations of its social constituency and the latter's impact upon the development of industrial technology. For illustrative purposes, there I have drawn primarily upon the communications industry which is the root of the modern electronics industry.

2. See Appendix for a view of the pre-war scale of the US's R&D system.

3. According to Pappademos (1983), the true extent of the post-Second World War military portion of the US's R&D spending is much higher than that indicated in the official data which I am using in Figure 1. His adjusted data show that, in 1980, instead of the 22% of the total R&D outlays which the official data gives to defense, 'a staggering 59.6% of the nation's scientific research and development was devoted to military purposes . . . since rising to 62.0 in 1982' (Pappademos, 1983, p.8). Most likely, these figures overestimate the magnitude of military participation, just as official figures tend to underestimate it. On the whole, however, there seems to be a better argument for suggesting underestimation rather than overestimation. Here I shall use the most conservative official figures to avoid any doubt.

4. For Mumford (1970), this was the inevitable result of the emergence of the new megamachine with the development of the atomic bomb. According to him, 'something worse than the invention of a deadly weapon had taken place: the act of making the bomb had hastened the assemblage of the new megamachine . . . in order to keep that megamachine in effective operation once the immediate military emergence was over, a permanent state of war became the condition for its survival and permanent expansion' (Mumford, p.256).

5. In the words of an OECD report (1971), after the war, 'There was no question as to the value of the products derived directly from military and space technologies such as jet aircraft, computers, and integrated circuits, although even in this area there was a question as to how much derived from direct government R&D investments and how much from induced demands for the products . . . But there is little doubt that, at least, their [these products'] development would have been considerably slower if the military incentive had been absent' (OECD, p.43). The same view is taken by Sundaram (1981), and by Salomon (1985) who questions the suggestion that public R&D investment would undoubtedly have a better effect on the economy if invested directly in the civilian sector; he asks, 'But could this happen in the United States if defence aims were not there to spur the administration to action?' (Salomon, p.85.)

6. For figures from 1930 to 1950, see US Department of Commerce (1965), Table 342.

7. Reagan assumed the US presidency in January 1981. 'The initial estimates showed Department of Defense budgets rising by $181 billion more than Carter administration's projections in the 1982 to 1986 fiscal years, a total of some 1.5 trillion. By the time the fiscal year 1985 budget was being proposed, the figure had risen again: for the years 1984-1988, Defense Department budgets were slated to exceed $1.8 trillion . . . In fact, the Vietnam build up cost only one-third of the Reagan rearmament in terms of constant dollars, comparing the 1965-70 period with 1981-1986.' (Tirman, 1984a, pp.ix-x).

8. Caspar Weinberger (Secretary of Defense), *Annual Report to the Congress. Fiscal Year 1986*, see Table 1 in the report.

9. 'From the early build up in production for European needs to the high point of wartime production, spending for defense jumped from less than 2 percent of gross national product (GNP) to nearly 40 percent' (Reppy, 1983, p.22). For yearly figures since 1940, see US Department of Commerce. Bureau of Census (1975).

10. See US Department of Commerce. Bureau of Census (1975). Series Y 472-87. In addition, by 1947 the total Federal outlays had fallen to approximately 40 percent from their highest wartime level.

11. For instance, the Berlin Blockade of 1948, the Chinese Revolution of 1949 and the first Soviet atomic bomb test in the same year.

12. 'The priorities of the 1945-70 period were largely determined by the Cold War. Government support for aircraft, nuclear and electronics R&D was both massive and effective. Firms in these industries became part of a special military-industrial complex, in which state-supported innovation was normal' (Freeman, 1974, p.41).

13. 'The geographically contained Korean War, which began on June 25, 1950, caused the United States not only to remobilize its conventional army, air and naval forces, but also to undertake urgent technological developments for the maintenance of a power base for worldwide stability' (Emme, 1967, p.581).

14. In Norman's words, 'Major weapons laboratories were established in the forties and fifties, links were forged between government agencies and private corporations as industry began to build new weapons and conduct military research under government contracts, and prominent scientists were pressed into service to advise defence agencies on their weapons programs' (Norman, 1981, p.73).

15. Other statistical sources suggest a 40% increase in industrial R&D expenditure for the period 1953-5 and a 100% increase for the period 1950-5. The same source suggests a 40% increase in university R&D expenditure for the years 1950-5. See US Department of Commerce (1960), Table 706.

16. For these and the data that follow, see NSF (1971).

17. 'The other side of the coin was an increase of government spending from $17.5 billion to a maximum of $103.1 billion in 1944' (Baran and Sweezy, 1975, p.162).

18. In 1948, Government expenditures had fallen to about one-third of its 1944 level while military expenditures had fallen to about 15%. See US Department of Commerce (1965), Table 342. The impact of these reductions on the industry heavily committed to wartime needs was enormous. Observe, for instance, what happened in the US aircraft industry. 'By the end of 1945 contracts amounting to well over $21 billion had been cancelled, and sixteen airframe plants remained in operation out of 66 that had been functioning a year earlier. It was unquestionably

a difficult time; between 1944 and 1947 the industry's sales dropped by over 90%, with corresponding effects on earnings and employment' (Rae, 1968, p.173).

19. In Birr's words, 'The imperatives of the Cold War in the 1950s and the race for national supremacy in space in the 1960s served to stimulate the demand for R & D activities in the post-World War II era' (Birr, 1979, p.202). We should also remember that it was in 1961 when President Eisenhower warned the US about the political danger of the growing military-industrial complex. 'This conjunction of an immense military establishment and a large arms industry is new in the American experience . . . In the councils of government one must guard against the acquisition of unwarranted influence by the MIC. The potential for the disastrous use of misplaced power exists and will persist . . . Akin to, and largely responsible for this has been the technological revolution. The prospect of domination of the nation's scholars by Federal employment, project allocation and the power of money is ever present but there is an equal and opposite danger that public policy could itself become the captive of a scientific-technological elite' (quoted by Pavitt and Worboys, 1977, p.27).

20. After the war the pressures on international economic competition were minimal since the destruction of Europe and Japan effectively left the US as the unchallenged economic power within the capitalist world. In particular, the country's technological strength had been considerably increased by wartime government support and by the immigration of many leading European scientists during the thirties and forties (Norman, 1979). As Vigier (1980) describes, 'The War which came to an end in 1945 was historically unique, in that, when peace finally came, nearly all the antagonists were broken and exhausted. The United States was the only exception. Its strength had more than doubled' (Vigier, p.150).

21. 'The launching in 1957 of the Soviet Sputnik satellite sent shock waves through the US government, for it suggested that the Soviet Union had pulled ahead in a key area of science and technology that had obvious military implications' (Norman, 1981, p.77).

22. Of course, not all space R & D activities had direct military purposes. The race to the moon, for instance, was primarily for purposes of international politics and to a lesser extent scientific research. The technological advances behind the race, however, had, like the Sputnik, clear military implications. In all, it is estimated that about 40% to 50% of the NASA activities are for military purposes [OECD (1968), Pappademos (1985)].

23. According to Reppy (1983), the postwar interlocking of military and industrial interests was the result of both a political choice and a technological necessity. In the latter respect, she identifies 'the significant increase in the rate of technological change in weapons and associated systems during and after World War II' (Reppy, p.23) as a major cause. In-house military facilities were perceived as not well suited to cope with the technological demands of rapidly advancing weapons systems. Thus, 'Technology and a political preference for private enterprise combined to

shift resources from the in-house establishment to outside contractors . . . currently government arsenals still retain a significant role in the production of munitions. Overall, 70 percent of DOD's programme for research and development is now performed by private industry, and the figure for procurement approaches 100 percent' (Reppy, p.23).

24. 'The United States began experiencing both inflation and balance-of-payments problems, and there is evidence of deceleration of productivity growth at about that time' (Nelson, 1979, p.50).

25. Between 1974 and the early 1980s, the situation of the US economy has been described as follows: 'the nation's economic performance reached several post-Depression records for unemployment, business failures, inflation, high interests rates, idle capacity, and just about every other category of significance.' (Tirman, 1984, p.14).

26. For these and other data given below in relation to the relative decline of the US economy, see Bueche (1979), Norman (1981), Fitt (1980), US Department of Commerce (1983), Tirman (1984), Thurow (1985). For various tables showing, through different economic indicators, the post-Second World War US economic performance in relation to that of other developed countries, see Niosi and Faucher (1985).

27. Among the negative consequences, it has been argued that 'the recurrent price inflation experienced in the 1970s was the result of the precipitous rise in military spending during the Vietnam War. President Johnson was unwilling to raise taxes to pay for the Indochina conflict, and the economic consequences were harsh' (Tirman, 1984a, pp.ix-x).

28. 'During the 1960s, a popular movement to end the war achieved a quite unprecedented scale. The resistance of American youth to the war against Vietnam—and Laos and Cambodia—has no parallel' (Chomsky, 1980, p.9).

29. 'The long campaign in Indochina had employed a dazzling array of technologies in support of the American cause. Although the entire arsenal was not exhausted . . . the sheer quantities and varieties of hardware used (and lost) in Southeast Asia brought home the military's increasing dependence on technology. As the nation's first "television war", Vietnam also displayed this employment of gadgetry to its large living room audience. Moreover, since the US expedition in Indochina was a failure, the pervasive use of technology was all the more intriguing' (Tirman, 1984, p.14).

30. Among the most conspicuous new attitudes and demands were 'rising concern about environmental degradation, growing alienation of workers stuck in boring, mindless jobs, a shift in aggregate demand away from material goods toward services, and increasing calls for the regulation of food, drugs, chemicals in the workplace, and other threats to health. Even economic growth itself, the very basis of industrial advancement for generations, came under assault as the prime cause of many of the problems facing industrial society' (Norman, 1981, p.56). See also Salomon (1985).

31. The political response to the new technological demands of society was epitomised in the New Technology Opportunities Program announced by the Nixon administration. 'Top research priorities, it was announced, would include clean energy, the control of natural disasters, transportation, emergency health care, and drug control. Symbolic of the whole approach was Nixon's announcement of the inauguration of a ten-year "war on cancer", an effort which, he confidently announced, would show that man was capable of using the same scientific skills that had taken him to the moon to come up with a cure for the most threatening of all human diseases' (Dickson, 1984, pp. 29-30).

32. See also data given in note 7 above.

33. Remember that the historic fall in the rate of profit which has characterised the crises of the mode of capital accumulation since the late 1960s has not just been associated with increased international competition but also saturating markets for the growth industries of the 1950s and 1960s and increased costs of labour and energy. Hence, the current technological and industrial challenge, although focused on international economic competition, relates also to the opening and expansion of markets and the search for productivity increases to counteract the effects of the costs and the power of labour.

34. Battelle-Colombus (1984) shows that, in real terms, R&D expenditure performed by industry began to rise in 1977 and has grown swiftly since with an estimated increase of 55% by 1985.

35. In practice, this dynamic has manifested itself most clearly in the great importance attached by all parties to what is known as the government-industry-university linkages. I shall elaborate on this topic below in relation to the development of microtechnology.

36. According to Smith (1983) in nineteen out of twenty leading military technologies the US is ahead of the Soviet Union.

37. For figures showing the trade-balance performance of high- and low- technology industries since 1960, see NSB (1981).

38. For a discussion concerning the differences in the civilian and military sector demands, see Tirman (1984), Gansler (1980), Kaldor (1985), Melman (1986).

39. During the three decades between 1950 and 1979, the US has spent an average of 8.4% of her GDP in defense. In contrast, Japan and West Germany have spent an average of 0.9% and 4% of their respective GDPS during the same years (US Department of Commerce, 1983).

40. As Schnee (1978) has put it, 'we should not expect the technological impacts of public programs to produce economic returns in the short run. The process of converting a significant technological advance into new products or processes is complex and time consuming' (Schnee, p.21).

41. According to Rosenberg (1983), 'It is doubtful that such data constitute good evidence of the changing relative pace of technological progress. It is more plausible to argue that the rising percentage of

patents in the United States obtained by foreigners has been dominated by commercial judgements and considerations such as changes in the size of specific markets, in the composition of demand, in relative prices, and so on. They do not necessarily indicate changing technological capabilities' (Rosenberg, p.282). There is evidence in the electronics industry to support Rosenberg's contention. Thus, Shapley (1978) reported that in some key areas of electronics, 'Americans have remained inventive, but they have been using the patent system less and less as a method of protecting their work' (Shapley, p.848).

42. For a recent analysis of the US defense industry, see Adams (1982). See also Gansler (1980).

43. It should be pointed out that although in the field of microelectronics the military have not favoured standardisation, this does not mean that they are in principle opposed to it. Indeed, the highly regimented nature of military organization often works towards standardisation, for instance, in such technologies as uniforms, rifles and combat gears [see Mumford (1967), Smith (1985)]. Thus, it may be suggested that the military will tend to promote standardisation in those areas where the technological frontier is moving slowly and their demands are massive, while in those areas such as microelectronics where the frontier is moving constantly and rapidly they will tend to promote the latest device for the specific needs they wish to satisfy.

44. It is interesting to note that the spin-offs argument has also been criticised from a political and moral point of view. Kulish (1976), for instance, sees it as part of the ideological arsenal which 'justify the arms race and glorify war as a necessary evil which makes it possible to consolidate good' (Kulish, p.55). Also, Thee (1981) has argued that 'if the tremendous resources wasted on military R&D could be channelled into peaceful applications, the accrued benefit to humanity would be equivalent to thousands of spin-offs from military technology' (Thee, p.57). Finally, given the essentially non-democratic nature of the military, Dumas (1984) has warned against the possibility of democracy being corrupted via technology transfer. 'To the extent that technologies that originated in service of the military bear the inherent values of that system, care must be taken when applying such technologies in the civilian sphere to avoid the subtle corruption of those very ideals central to democratic society as a whole' (Dumas, pp.145-6).

Notes to Chapter 3

1. According to Brooks (1976), the greatest contribution of Bell Labs to victory was in the field of radar, where the Bell system shared the leadership with the MIT's Radiation Laboratory. Before the war was over, the system 'had produced, from Bell Labs designs, 57,000 radar units of seventy different types for airborne, ground, and naval use, constituting about half of all US radar manufactured' (Brooks, p.211). In addition, the total absorption of the Bell system into the war effort was enormous. Thus,

Western Electric, its manufacturing arm, was 54% converted to war work in 1942 and by 1944 the figure reached a peak of 85%. On the research front, Bell's commitment to military work was complete. 'Beginning in 1942, virtually all the Labs' six thousand people were engaged in such a work' (*ibid*., p.210).

2. At the same time, the government made sure that labour costs would be kept from rising on account of the military emergency. 'Throughout the war, for the labor force as a whole, wages were frozen at 15 percent above 1941 levels . . . while prices rose 45 percent and profits increased 250 percent' (Noble, 1984, p.22).

3. This project eventually came to involve one-third of the electronics industry at a cost of approximately $1 billion. As a result, manufacturing capacity for tubes, capacitors and resistors was increased tremendously (*Electronics,* 1980).

4. It was built by the Moore School of Electrical Engineering at the University of Pennsylvania for the US Army. In fact, the Moore School had been working with the Army (Ballistic Research Laboratory) on ballistic problems since the 1930s. War only made the need for results all the more urgent. In Soma's view, 'the military had a huge demand for computational power for the development of ballistic firing tables. This demand was intensified by the occurrence of World War II' (Soma, 1978, pp.1-2). One of ENIAC's first practical uses was in the atomic-bomb project at Los Alamos Laboratory in New Mexico. See Fleck (1973) and *Electronics* (1980).

5. During the war, 'The speed and maneuvrability of aircraft quickly made traditional gun-laying techniques obsolete. In order to meet the pressing military requirements, considerable numbers of highly capable scientists went to work on the problem of designing and constructing automatic position control, or servomechanisms systems' (Diebold, 1952, p.19).

6. The progress we have described was not all that came out of the wartime years. There were also major advances in radio, TV and instrumentation. Among other devices, magnetic tape was produced along with recording heads and tape recorders for sound movies and radios. See *Electronics* (1980) and Atherton (1984).

7. There are other aspects of a more indirect nature such as taxation, legal dispositions on competition, and educational policy which undoubtedly have played an important part in the global approach to the industries concerned. For our purposes, we need only to select those aspects which are the most revealing.

8. 'The new high-performance, high-speed aircraft demanded a great deal of difficult and expensive machining to produce airfoils (wing surfaces, jet engine blades), integrally stiffened wing sections for greater tensile strength and less weight, and variable thickness skins' (Noble, 1979, pp.24-5).

9. According to Sciberras and Payne (1984), early NC machines were also supplied by 'Bendix Aviation Corporation (with an in-house built machine and controls) and Jones & Lamson's NC turret lathe' (Sciberras and Payne, p.39).

10. Between 1949 and 1959, the military spent at least $62 million on the research, development and transfer of NC (Noble, 1979). See also Kaplinsky (1984).

11. 'Government supported R&D programs were both antecedent and directly related to first- and second-"generation" commercial machines as well as to significant intragenerational improvements' (Katz and Phillips, 1982, p.166).

12. 'The cadre of people working on computer development came from universities, various government departments, and industry. They had frequent formal and informal contacts with one another' (*ibid.*, p.167).

13. 'IBM, Remington Rand, NRC and Borroughs were all obvious companies to develop computer capabilities because of their stake in the business machine market, which was likely to be displaced by the new technology' (Brock, 1975, p.13). It was only in 1959 that companies based entirely upon computers successfully entered the field. These were Control Data Corporation (CDC) and Digital Equipment Corporation (DEC). 'Each of the previous entrants . . . was an established large corporation in other lines of business before entering computers' (*ibid.*, p.15).

14. The SAGE System 'was set up to protect the United States against a surprise attack. SAGE consisted of a network of radar-fed computers that continuously analyzed every cubic foot of air space around the US, instantly tracked all approaching aircraft, and decided on an appropriate response . . . The original computer programs for SAGE consumed 1800 man-years effort; the total system cost has been placed at close to $2 billion' (Schnee, 1978, p.16).

15. 'Watson, Jr. suggested this might be an opportunity to become involved with government agencies as supplier for electronic computers . . . [In his view] large-scale electronic computers might be crucial to the company's future' (Pugh, 1984, p.29).

16. For a brief description of mergers, joint ventures and outright sales up to the mid 1960s in the computer industry, see Sharpe (1969).

17. This renewed space-defense drive for computers was instrumental in the emergence of Control Data which in 1971 ranked second to IBM. In Schnee's description, 'Space-defence business provided the initial impetus for Control Data's growth; almost all of the company's pre-1960's sales were to government agencies' (Schnee, 1978, p.13).

18. Apart from the work at Purdue University in the late 1940s, which came close to inventing the point-contact transistor before Bell Labs, in the fifties and sixties 'the aggressive development policy of Stanford University, together with the strength of the physics and electrical engineering faculties at Stanford and the nearby University of California at Berkeley, was

responsible for the location of numerous merchant semiconductor firms in the Santa Clara Valley. University-industrial ties in the Bay Area and in the Boston area were especially close in these early years, and many new ventures involved university faculty in important consulting or managerial roles' (Levin, 1982, p.47).

19. Also, one has to remember that work on semiconductors at Bell Labs had been interrupted by the war. In 1936 it had started being pursued by Shockley and Brattain, but as Brooks explains, 'early in 1940 the two scientists' energies were diverted into war work. The coming of the war thus postponed the coming of the basis of postwar electronic technology' (Brooks, 1976, p.203). See also Braun and MacDonald (1978).

20. According to Brooks (1976), Bell Labs grew into a giant with the Second World War and has since developed into a supergiant. By the mid 1970s, it maintained seventeen locations in nine states, and an eighteenth in Kwajalein Atoll in the Pacific Ocean. It had about 16,500 employees of whom 44% were professionals and 23% technical assistants; these included more than two thousand Ph.D holders and almost four thousand with master's degrees. The total Bell's budget for the year 1974 was $625 million. About a decade later, just before AT&T's divestiture in 1984 (see below), Bell Labs had 3,000 Ph.Ds and more than 26,000 employees in more than a dozen locations. Its budget had grown up to $2 billion (Wallich, 1985).

21. In January 1956 a Consent Decree enabled the Bell System to become a regulated monopoly, thus avoiding its break up. In exchange, its 'two substantive provisions restricted the scope of Bell System activities and required liberal licencing of Bell System patents' (Brock, 1981, p.191).

22. Just consider that, by 1950, as one observer put it, 'Almost 90 per cent of the semiconductors items . . . in commercial production came right out of *Mother Bell's Cookbook*' (Silk, 1960, p.75).

23. One reason for this is that the Bell System has acquired most of the important patents (Tilton, 1971). Another reason is that 'Once the critical patents were freely available, the cumulative nature of technical progress in the industry guaranteed that patents would either be widely cross-licenced or simply ignored' (Levin, 1982, p.81). It is interesting to note that the fundamental concept of liberal licencing has lasted for several decades and only recently came under serious challenge as a result of the fierce competitive battle being waged by the US and Japanese semiconductor industries (Molina, 1986).

24. Table 2 does not show production support for integrated circuits but it is estimated that $10 million were given to the semiconductor industry as early as 1959 to build a production capability for integrated circuits (Linvill and Hogan, 1977).

25. The production refinement contracts for transistors called for federal support of all engineering design and development effort, while the firms involved paid for capital equipment and plant space (Levin, 1982).

26. At this time, the older companies were themselves providing about two-thirds of their R&D expenditure and the new firms about 90% (Braun and MacDonald, 1978).

27. Gordon Teal, whose work in crystal-growing techniques at Bell much advanced the production process of transistors, moved to TI in 1952. By that time, Teal was working on single-crystal silicon because of the expected high-temperature capability (Teal, 1976) which was so much in demand by the military. TI had made it its goal to be the first to make a silicon transistor available to the military (Levin, 1982). According to Teal (1976), 'It is hard to overestimate the impact of TI's commercial lead in silicon transistors' (Teal, p.637).

28. The IC was simultaneously invented by Fairchild, the company which also developed the planar process so crucial for the mass-production of semiconductor devices. Fairchild did not receive direct R&D military support and the development of the planar process was primarily related to the desire to improve the production process of discrete transistors [OECD (1968b), Mowery (1983)]. The IC itself was developed by Fairchild more in response to TI's impending announcement [Wolff (1976), *Electronics* (1980)] but in 1958 the company did receive a USAF's $1.5 million contract to supply silicon diffused transistors. According to Levin (1982), this helped the company to further development work on the planar process. Later, in the early sixties, Fairchild was to benefit greatly from the space race demands for integrated circuits.

29. The NASA Microelectronic Reliability programme set criteria for the acceptance of all semiconductor devices and also required the inspection of all production facilities under contract to NASA (Wilson *et al.*, 1980).

30. Some of the important R&D work backed by the military was on MOS, CMOS and Silicon-on-Sapphire technologies. RCA received funding for these three programmes (Wilson *et al.*, 1980).

31. Prior to 1955, the military demand was perhaps even higher. For instance, of the 90,000 transistors produced in 1952, mostly point contact devices from Western Electric, the military bought nearly all (Braun and MacDonald, 1978). At the time, however, industry had also begun to incorporate the transistor. RCA, for instance, demonstrated an all-transistor TV in 1952 and Bell had started to use it in the telephone system in 1951 (MacDonald *et al.*, 1981).

32. A new variety of transistor, for example, would command a very high price when it first came onto the market if it had unique advantages. The military was prepared to pay exorbitant prices to have the very best as soon as possible (MacDonald *et al.*, 1981).

33. For a discussion of how the Minuteman Missile programme shaped the development of early integrated circuits, see Platzek and Kilby (1964).

34. For instance, 'the improved Minuteman program called for the development, design, fabrication, and delivery of twenty new types of semiconductor integrated circuit in a six-months period. Although this

number was subsequently reduced to eighteen types, it undoubtedly exceeded the total number of types in production in the industry in mid 1952' (Platzek and Kilby, 1964, p.1678).

35. 'Initial defense demand for components, the rapid expansion and the size of demand all led to rapid dynamic or "learning" economies and laid the base for subsequent price reductions. All the means which encouraged entry into the industry . . . also led to a more competitive situation and lower prices. Cost reductions and price competition engendered by the military probably resulted in much more rapid penetration of industrial and consumer markets by transistors and integrated circuits than would otherwise have been the case' (Utterback and Murray, 1977, p.4).

36. 'Antitrust policies apparently shifted to some extent the market positions and towards manufacturing for internal use and generating revenues from innovative efforts in their other markets . . . Some of the older firms like Sylvania, Sperry Rand, Westinghouse and General Electric eventually had to drop out of the so-called merchant market altogether, although they continued to manufacture for themselves' (Wilson *et al.*, 1980, p.157).

37. See Utterback and Murray (1977), Golding (1971), Freeman (1974), Tilton (1971), Braun (1980), Schnee (1978).

38. I shall take only the case of semiconductors to illustrate the point, but it seems plausible to assume that for the other microtechnology industries the pattern is broadly the same as for the semiconductor industry. Indeed, as seen in Figure 13, for the case of computers, already in 1970 the government's share in the total value of general-purpose computers was about 15%.

39. For instance, industrial leaders first introduced microprocessor units for commercial markets. 'The Fairchild F8 and the Texas Instruments TMS 9900 were followed only later by military devices, such as the Texas SBP 9900' (Sciberras, 1980, p.289).

Notes to Chapter 4

1. As one observer put it in relation to the university, the focus of the Vietnam protest movement of the early 1970s, 'The great crisis of ideology during Vietnam has all but evaporated among faculty, student and staff' (R. Sproull, president of the University of Rochester, quoted by Walsh (1981), p.1003).

2. The primary source of university research funds for the past twenty-five years has been the federal government (Stanfield, 1980). In Dickson's words, 'federal support for education and research increased in the 1950s and 1960s much faster than that for industry. As a result even though industry sponsorship of research continued to grow significantly in absolute terms, as a *proportion of* academic research funding it dropped precipitously—from 10 percent of the total in 1955 to 5 percent in 1960, and less than 3 percent in 1970' (Dickson, 1984, p.64). See also Pappademos (1983).

3. For a discussion about the specific benefits accruing to industry, the university and the country from a closer interrelation between these two institutions, see Prager and Ommen (1980), David (1979) and Nelkin and Nelson (1987).

4. According to Noble (1982), this complementarity of interests has developed into a much deeper integration of interests spanning social and political views and actions. This has taken place, primarily, through the merging of interests of corporate capital and what Noble calls corporate academics. 'Quite naturally, and without the need for any conspiracy, their two identities merge in their thought and actions. Thus do academic interests, and often personal and professional interests as well, converge with corporate interests. These corporate academics have come to share the world view of their confreres in the board room' (Noble, p.147). See also Dickson (1984) for a similar analysis.

5. For instance, as two analysts have described, NASA and DOD, 'through their major technology development and procurement programs, have stimulated the formation of research consortia to direct their collective capabilities to the solution of specific technical problems. The Department of Energy has stimulated university-industry-government cooperation in R&D related to specific energy technologies . . . The Department of Commerce enlists the aid of universities in facilitating the introduction and application of technologies designed to improve the competitiveness of industries in the international market place' (Prager and Ommen, 1980, p.382).

6. The aim of the centres was the sponsoring of 'joint industry/academe research on generic technologies, in which individual firms have little incentive to invest, but which may have significant economic or strategic importance, such as manufacturing technology. The centers are also to provide assistance and advice to individuals and industries, particularly small businesses' (*Physics Today,* December 1980, p.55).

7. DARPA was formed in the late 1950s in the wake of the technological race unleashed by the Sputnik. Its aim has been to promote basic research and until recently its role, particularly of its Information Processing Techniques Office (IPTO) has been considered of great importance for the development of computer science in the US. As Ornstein *et al.* (1984) explain, IPTO 'headed by distinguished computer scientists, has established itself as the principal government sponsor of computer research at universities and industrial laboratories. Much of this research has been generic in nature — applicable to a large variety of military and non-military problems' (Ornstein *et al.*, 1980, p.11). As we shall see later on, all this has begun to change as a result of the renewed military drive on technology.

8. Centres appeared in other technological fields as well, for instance, biotechnology, building technologies, and polymers. See Ploch (1983) for a list of centres and their research activities, including those in the microtechnology field. See also Nelkin and Nelson (1987).

9. Various authors have described these centres. See Ploch (1983), Botkin *et al.* (1982), Norman (1982), Mason (1980).

10. Among those sponsoring CIS in 1983 were GE, TRW, Hewlett-Packard, Northrop, Xerox, TI, Fairchild, Honeywell, IBM, Tektronics, DEC, Intel, ITT, GTE, Motorola, United Technologies and Mosanto. In sum, many of the companies dominating the electronics industry (Norman, 1982).

11. RPI, for instance, has brought together faculty members of five different departments for its Center for Integrated Electronics. Likewise, the Robotics Institute at CMU draws on faculty from several different departments, especially computer science and mechanical engineering (Norman, 1982).

12. 'Planning for the program began in 1978 immediately after military intelligence reports revealed that the US advance in the electronics embodied in field weapons systems had been significantly eroded' (Levin, 1982a, p.49). See also Perry and Roberts (1982) and Weisberg (1978). Classified intelligence data showed that the US technological lead had slipped from an estimated five to ten years to three to five years and was continuing to diminish (Connolly, 1978).

13. 'The goal of the program is pilot production in 1986 of processors containing 250,000 gates, operating at clock speeds of at least 25 MHz, and performing several million to several billion operations per second' (Sumney, 1980, p.24).

14. As the VLSI project, the Fifth Generation Project is in fact part of a Japanese strategy to gain international superiority in the science-based industries. This goal is explicitly asserted in *Vision of MITI's Policies in the 1980's*. 'It is extremely important for Japan to make the most of her brain resources, which may well be called the nation's only resources, and thereby develop creative technologies of its own . . . Possession of her own technology will help Japan to *maintain and develop her industries' international superiority* and to form a foundation for the long-term development of the economy and society', (quoted by De Grasse, 1984, p.95). In this context, the Fifth Generation Project is specifically envisaged as a) increasing productivity in low-productivity areas particularly white-collar work; b) providing intelligent assistance to managers with high-level expertise necessary for making important decisions; c) developing new computer technology based on Japanese conceptions, thus breaking with the dynamic of following the computer lead of other countries; d) saving of energy and resources through better information and its optimal use; e) helping Japan to cope with the problems of an aging society, for instance, in health management; f) helping international cooperation through the development of translation and interpreting systems; g) expanding human abilities by amplifying, not merely physical labour, but intelligence too; and, finally, as part of the general development of knowledge industries, the Fifth Generation Project is seen as helping in the promotion of stable, consistent, and sophisticated judgements in politics, administration, and

industry (Feigenbaum and McCorduck, 1983). More relevant to US-Japan competition, M. Dertouzos, director of MIT's Computer Laboratory, has seen the project as a 'carefully conceived blueprint of the research and engineering needed to leapfrog the US computer industry and destroy its world supremacy' (quoted by Marbarch *et al.*, 1985, p.60).

15. The second phase will be devoted to systems development, the third is one of gradual and sequential deployment and the fourth will be reached when ballistic missile defenses are fully in place. Offensive missiles are expected to be negotiated to a minimal amount.

16. For another explanation of the technical demands involved in the realisation of SDI, see Adam and Wallich (1985) and Fischetti (1985).

17. See Panofsky (1985), MacKenzie (1985), von Hippel (1985), Lin (1985), Lamb (1985), March (1985), Aftergood (1986), *Science for People* (1986), Enfield (1987) and for an overview of both sides of the argument Adam and Horgan (1985). In relation to the boycott organised by academic scientists against SDI in the US and the UK, see Shulman (1986) and *Nature* (30 October 1986).

18. 'Because the products of this industry are the crucial intermediate inputs of all final electronics systems, competition in the semiconductor industry will be at the center of competition in all industries which incorporate electronics into their products and production processes . . . The semiconductor industry is therefore strategically vital to the future growth of knowledge-intensive industrial development within the US economy. For the foreseeable future the relative economic strength of all advanced industrial economies will rest in part on their capacity to develop and apply semiconductor technology to product design and production processes. Thus the loss of leadership in this one industry would mean the loss of international competitiveness in many of the advanced technology sectors that have been the basis of a US advantage since the Second World War' (Borrus *et al.*, 1982, pp.1-2).

19. 'Contractors and subcontractors alike expect to switch their best engineering talent to the VHSIC project because it requires the most advanced skills . . . The most immediate impact will be to leave potentially large vacuums in nonmilitary research efforts, especially at highly experienced scientific levels. Yet it is precisely those skilled mature engineers who are most difficult to find and that industry most needs' (Botkin *et al.*, 1982, p.78). The same concern is raised by Weizenbaum (1985) as regards the impact of SDI on the field of AI. For a historical account of the way in which past involvement of the military in the development of US semiconductor technology has diverted resources away from civilian lines of pursuit, see Misa (1985). Also, the Appendix gives similar evidence in relation to the pre-Second World War development of US communications technology.

20. This judgement is supported, for instance, by Utterback and Murray's conclusion concerning the electronics industry: 'Major advances in semiconductor technology have with few exceptions been developed and

patented by firms or individuals without government research fundings. Far fewer patents have resulted from defense supported R&D than from commercially funded R&D, and a far smaller proportion of those which have resulted from defense support have had any commercial use. Few examples can be cited of systems or equipment designed for defense purposes and later directly transferred to commercial use. Those transfers which have occurred, for example of navigational radar, have required essentially new designs for commercial applications' (Utterback and Murray, 1977, p.2).

21. Details of the Japanese strategy are discussed by Borrus *et al.* (1983), De Grasse (1984), Sigurdson (1983), mainly in relation to the technical-economic context of Japan. More broadly, some commentators have drawn attention to the fact that, alongside the techno-economic and policy factors, cultural factors such as Japan's capacity for relatively rapid social organisational adaptation also play an important part [Freeman *et al.*, (1982), Kikuchi (1983)]. In a recent paper, Freeman (1985) has argued that changes in world technological leadership are associated with the emergence of new technological paradigms and the associated changes in industrial structure; and that it has been Japan which has shown most clearly the capacity to exploit the new information paradigm. For an analysis of the techno-economic paradigm based on microelectronics, see Perez (1985).

22. This is not to say that the Japanese electronics industry, particularly the large corporations dominating it, do not indulge in military projects. They do but so far on nothing like the scale of the military involvement of the US industry. Berger (1985) provides an idea of Japan's military programmes and the companies which are working on them. According to this author, the Japanese, who have spent a yearly average of $300 million to $400 million on US military equipment over the last five years, may now be preparing themselves to become a force in the international military market as well. He quotes a Japanese executive as saying, 'The question is no longer whether Japan will become a factor in the international military market but when' (Berger, 1985, p.34).

23. The same argument prevails as regards SCP and the benefits of the increased research funds for computers and AI. The fact that much of the research will be generic is seen as enabling the US to maintain its position, particularly in AI, while civilian spillovers will help to counteract the Japanese challenge. For instance, 'The same technology that can be used to simulate an antitank missile smashing into a heavily armoured tank can also be put to work on less martial arts' (Marbarch *et al.*, 1985, p.65).

24. Similarly, in the field of computers, in 1982 the Japanese companies Fujitsu, Hitachi and NEC announced supercomputers faster than the most powerful American machines at the time in the market, the Cray X-MP and Control Data's Cyber 205 (Marbarch *et al.*, 1985). Here, however, as we shall see later on, US supremacy is very strong, with IBM clearly dominating the world market. In terms of R&D, IBM alone had an annual budget of $1.5 billion in 1982. Of course, only a limited portion of this budget goes on long-term research projects. In contrast, the MITI's

budget for the long-term Fifth Generation Project has been announced at $450 million for a ten-year period and matching funds from participating companies will take it up to $850 million (Feigenbaum and McCormack, 1983). In turn, DARPA's budget for the US response to the Fifth Generation Project, SCP, started with $600 million for the first five years of the ten-year period envisaged for the plan. Thus, the Japanese are certainly spending less than the US in computer R&D. As with microelectronics, however, the impact of research expenditures is bound to depend on the direction it is channelled into.

Notes to Chapter 5

1. For instance, in 1980, more than half of the development time for the newest microprocessor was spent in software (Robinson, 1980).

2. According to Fisher *et al.* (1983), TI has marketed a broad range of computer equipment since the days of the transistor.

3. 'It is foreseeable that in the next few years large manufacturers using microelectronic components will find it hard to compete and survive without internally producing such components' (Lamborghini, 1982, pp.134-5).

4. This situation has now changed with the end of the regulatory controls in the early 1980s.

5. 'Despite the continuing restrictions of its 1956 consent decree . . . AT&T expanded its offering of computer-related products and services during 1964-1969. As it had done in the 1950s, AT&T competed in the computer industry in at least two ways in the 1960s. The first involved Western Electric's manufacture and marketing to the Bell System operating companies of stored-program-controlled electronic switching and automatic intercept systems . . . The second form of competition consisted of AT&T's offering of EDP [electronic data processing] products and services to non-Bell customers. While this business was also a large one, AT&T was restricted in the extent to which it could compete in this area by its 1956 consent decree. These restrictions did not apply, however, to sales to the US government, including the military, and AT&T was active in such sales' (Fisher *et al.*, 1983, pp.279-80).

6. Among the important provisions of the 1982 settlement, AT&T agreed to divest itself of its local operating companies early in 1984. In exchange, apart from the freedom to compete, AT&T was allowed to retain almost intact Bell Labs, its lucrative Long Lines division and its manufacturing arm Western Electric (*Financial Times,* 10 January 1983). For the events involved in the decision to divest, see Bell (1985). For an examination of Bell Labs after divestiture, see Wallich (1985) and the *Financial Times* (9 February 1987). For the record, before the divestiture took effect in January 1984, the Bell System was the largest corporation ever to have existed; its 'total assets were some $150 billion, making it bigger than General Motors, IBM, General Electric, US Steel, Eastman Kodak and Xerox combined. Its annual revenues were nearly $70 billion,

representing approximately 2 percent of the US gross national product. Its net income was nearly $6 billion, slightly smaller than the total budget of the National Aeronautics and Space Administration. It employed just under a million people . . .— making it the largest private employer in the country' (*ibid.*).

7. See *Datamation* (July 1982) special issue on IBM versus AT&T, Foremski (1984), Bylinsky (1984), Kirland (1984), *Financial Times* (18 January 1983), Kozma (1985). For the position and the competitive strategies of the two companies under the regulated period see Anderson (1977), Farber and Baran (1977).

8. According to Kirkland (1984), 'while AT&T mounts a risky and expensive attack on IBM's home ground — computer hardware and software — IBM will mostly avoid frontal assaults on markets dominated by the likes of AT&T, GTE, ITT, Northern Telecom and the regional Bell operating companies' (Kirkland, p.37).

9. Both the UNIX operating system and the ACS were developed in the late 1970s as AT&T saw the need to strengthen its computing capabilities for the coming competition. See Fisher *et al.* (1983) and also Fishman (1981).

10. The investment in MIC Communications Corp. is a recent move by IBM. Before that the company had a stake in Satellite Business Systems (SBS), an advanced satellite communications network for private-line service via small antennae on the customer roofs. The SBS venture was announced in 1974 when IBM also saw the need to expand into telecommunications [see Fishman (1981) and Anderson (1977)]. As part of the recent move, however, IBM sold its stake at SBS to MCI Communications while investing in the latter company. According to an observer, 'By investing in MCI and making it more viable, IBM's goal is to cut into AT&T's resources from selling services, so AT&T can't use those resources to help fund its attempts in the computer business' (Kozma, 1985, p.29).

11. These are Telenet, a subsidiary of General Telephone and Electronics (GTE) the biggest of the independent US telephone companies; Tymnet, part of Tymshare a computer service house; Graphnet, part of Graphic Scanning, a data communication company; and Uninet, part of United Telecommunications [*Financial Times* (18 January 1982), Webster and Robin (1979)]. Recently, GTE and United Telecommunications have merged their Telenet and Uninet data transmission operations, in a move that is expected to create the dominant carrier in the business (*Financial Times,* 23 January 1986).

12. Ford Motor Co., for instance, has its own microelectronics centre — Ford Microelectronics Inc. — to stay ahead in the race to use automotive semiconductors (*Electronics,* 26 January 1984). Also, Ford Aerospace & Communications Corp. is currently second only to General Motors in the supply of satellites (see Zorpette, 1985).

13. A similar pattern of high barriers of entry is found in the computer industry where, in 1978, R&D outlays as a percentage of sales was 6% and the capital requirements to enter the mainframe integrated systems market were estimated at $1 billion in the mid 1970s (Soma, 1978). In addition, in this industry account has to be taken of the fact that IBM, its dominant force, uses accelerated five year depreciation on electronic switches 'which means that it has effectively written off a computer . . . in two years' (Anderson, 1978, p.93).

14. 'Even the US market is no longer big enough to sustain a fully fledged information technology industry purely on domestic demand . . .[This] . . . industry is therefore characterized by its multinational nature as no single economy can sustain a full range of computer and electronics production' (Duncan, 1982, p.93).

15. It is a characteristic of the Japanese strategy to build up strong capabilities in targeted sectors such as semiconductors by using, among other means, an initially protected internal market as a platform for worldwide competition [Borrus *et al.* (1982), Merrifield (1983), Schmitt (1984), SIA (1981), Duncan (1982)]. Clearly in the case of the electronics industry, or of any industry for that matter, such a strategy can work only if the market is sufficiently big and dynamic to permit the formation of economies of scale.

16. Another area in which the Japanese have become a major force is that of consumer electronics where they have come to dominate the markets for colour televisions and video recorders. See Walsh (1977) and Fusefeld (1982). This is consistent with the importance of the consumer electronics market in Japan.

17. Because of its strategic importance US-Japan semiconductor competition has been the subject of various studies. See Borrus *et al.* (1982), SIA (1981), Robinson (1980b), Walsh (1982) and Ferguson (1983), NAE/NRC (1984).

18. In fact, early 1985 estimates still put TI and Motorola in first and second place in the league of top semiconductor companies (*Electronics Week*, 1 January 1985). In the meantime Europe's share of the world semiconductor market fell from 14.6% in 1976 to 9.5% in 1983. (*The Economist,* 8 December 1984).

19. The recent devaluation of the yen against the dollar has almost fulfilled this prediction. In 1986, although in volume terms Japanese companies' chip sales remained at the same level as in 1985, in value terms their share of the world chip market increased to 48% with the US share falling to 40% (*The Economist*, 7 March 1987). For the same reason, the Japanese semiconductor market has now become the largest in the world, clearly surpassing that of the US. A recent market report puts the Japanese market at $16.4 billion in 1987, while the US market for the same year was $13.1 billion (*Electronics,* 7 January 1988).

20. It is important to note that it is in the mass memory market that the Japanese success has been most outstanding. Thus, in 1981, they captured 70% of the 64k RAM's world market (*The Economist,* 19 June 1982). On the other hand, in microprocessors, where there is an important input in software, the Japanese challenge has been less effective. Thus in 1980, *The Economist* stated that, 'In microprocessors, the Japanese are not yet challengers . . . They lack the software skills. That could change, but not overnight' (*The Economist,* 8 November 1980, p.93). In 1982, however, *Newsweek* reported that 'Japan still imports most of its logic chips — microprocessors — but the overall lead once enjoyed by American makers has eroded (*Newsweek,* 9 August 1982, p.23). And, in 1984, a report on the Japanese firm NEC put this company second only to Intel (US) in worldwide sales of microprocessors. NEC also had captured 40% of the Japanese microprocessors market (*The Economist,* 15 December 1984).

21. 'This time the Japanese are winning business from US manufacturers of chipmaking machinery, and more is at stake than the $5.8 billion-a-year semiconductor market. The country with the best hardware may gain a technological edge in chips' (Uttal, 1984, p.58).

22. According to Inaba (1983), Japanese makers of semiconductor equipment 'are gearing up to do more business in the US not only to increase production of their equipment, but to obtain technological feedback from US users . . . So far, the Japanese have a very small share of the equipment market in the US. Nevertheless, they count on equipment that has a price/performance edge over US equipment to expand their market presence' (Inaba, p.10).

23. Fanuc has as one of its main shareholders Fujitsu Ltd., the leading Japanese computer company . Fanuc developed out of the latter company and was established as an independent company in 1972. Its original name was Fujitsu Fanuc Ltd. but later this became simply Fanuc Ltd. In 1975, Siemens (Germany) also became a main shareholder of Fanuc as both companies concluded a mutual assistance agreement (*Introduction to Fanuc,* published by the company, n.d.).

24. See also Sciberras and Payne (1984). Technologically, some observers contend that the Japanese are now significantly ahead of the US in machine tool control technology and have real advantages in the technology of high-volume production. Others, however, such as the US National Machine Tool Builders Association consider the difference to be small (NAE/NCR, 1983).

25. IBM became the leader of the CAD/CAM market only in 1984. Until then, Computervision was the dominant supplier (*Financial Times*, 16 July 1985).

26. A report suggests that in 1984 the Japanese company Fujitsu had taken fourth place (*Business Week,* 16 July 1984). This is now the case after the merger of Sperry and Burroughs, the US computer companies holding third and fourth places respectively. More recently, a joint venture

involving NEC (Japan), Honeywell (US) and Bull (France) has been proposed (*Financial Times*, 25 September 1986). These developments are reshaping the leadership of the computer market although IBM's position remains very strong.

27. 'Between 1975 and 1980 annual production of Japanese computer manufacturers increased by 139 percent to $4,700 million . . . Exports as a share of production increased from 6.8 percent to 10.7 percent ' (Gregory, 1983, p.28).

28. Others even thought that IBM would be overtaken by the Japanese (*The Economist,* 19 June 1982).

29. It is not that the Japanese are weak at producing software. As one commentator puts it, 'Japan's relative shortage of software is not due to an inherent inability to write programmes. It is mainly a matter of coming later into the field and not having built up as large a stock of programmes as Western companies have. So it will take time, perhaps until the 1990s, for Japan to accumulate a comparable software base' (*The Economist*, 19 June 1982). A similar view is stated in a *Business Week* report in 1984 on the software industry. 'Even with MITI's help, it will be a long time before Japan's software industry catches up with its US competitors — if it ever does. In 1982, the latest year for which figures are available software sales in Japan were only $1.4 billion — just one-quarter of the total US software sales that year' (*Business Week,* 27 February 1984, p.71).

30. NEC adopted a more independent strategy developing its own operating system and its own international marketing organisations with an emphasis on personal computers. NEC's total computer sales in 1981 surpassed those of Hitachi to become second only to Fujitsu among Japanese computer makers (Gregory, 1983).

31. Indeed, Hitachi and Fujitsu have increased software research expenditure to about a third of their total R&D effort. Also some companies like Toshiba are building software factories. Toshiba has recently completed a factory that employs 3,000 software engineers and it is building another that will employ 2,000 more programmers. In addition, MITI is setting up several research laboratories to work on software (*Business Week,* 27 February 1984). On its part, NEC has made software-development a priority in its computer and communications systems laboratory which accounts for one quarter of all the company's research. In the late 1970s NEC had established a software production technology laboratory and, in 1984, it was said that '25% of NEC's total research staff is engaged in software-related work' (Berger, 1984, p.36). More generally, Berger (1984) reports that in the aftermath of the IBM-Hitachi computer-software affair, an R&D boom among Japanese electronics companies has taken place. 'The settlements accepted by Hitachi and other IBM-compatible makers in Japan . . . has shown firms . . . that independent product development is a safer path' (Berger, p.34).

32. 'Japan's interest in AT&T is understood to be as an alternative to IBM, which successfully "stung" Hitachi and other Japanese manufacturers in

1982. MITI is concerned about the future of the Japanese computer efforts given the relatively unproductive Japanese software industry' (*Datamation*, 1 October 1984, p.64).

33. In relation to AT&T and IBM, for instance, Foremski (1984) says that, 'In spite of the two companies' huge individual resources in terms of assets and research and development efforts, neither of them has the ability to develop the necessary technology and to accumulate expertise in a short enough time' (Foremski, p.16).

34. 'These firms belong to *kieratsu* — large families of companies that include robotics, semiconductor, and other high technology firms. *Kieratsu* have ready access to relatively inexpensive, long-term financing, frequently through special relationships with banks, so member firms are able to plan large purchases of capital equipment. The *kieratsu* are better prepared than US producers to make optimal decisions regarding technology and strategy' (Ferguson, 1983, p.28).

35. In 1984, *The Economist* asked. Why can NEC withstand the slower market growth that is now dragging down American semiconductor makers? Its answer was that 'One reason is that Japanese demand for chips has been less volatile than America's. The four top Japanese chipmakers — NEC, Fujitsu, Hitachi and Toshiba — are four of the biggest chip consumers too. This makes demand easier for NEC to forecast, and it moderates the industry's boom-bust cycles' (*The Economist*, 15 December 1984, p.73).

36. According to a report, 'This increased collaboration reflects companies' fears that if they do not offer complete product lines at competitive prices, the Japanese . . . will cut deeply into the information processing business . . . Also worrying these companies is the entry by International Business Machines Corp. into almost every segment of the market' (*Business Week*, 16 July 1984, p.49).

37. Some of the larger companies 'are looking for partners that can provide them with the expertise such as the communications and software needed to tie all the pieces together . . . the smaller companies are seeking greater financial strength from their new relationships so they can continue to develop better technology and expand their manufacturing facilities' (*Business Week*, 16 July 1984, p.50).

38. See also Hazewindus (1982) and *Business Week* (3 December 1979).

39. As Kozma (1985) has indicated, 'To shore up its position and to fill the gaps in product lines, it [AT&T] is forming strategic alliances with anyone and everyone who is interested in UNIX, the Bell Laboratories operating system that AT&T is championing as the software standard for many uses' (Kozma, p.26).

40. A different report puts the figure of 146 acquisitions of companies only within the software industry (*Business Week*, 27 February 1984). In addition, it says that 'To provide a full complement of software for their computers, makers such as Hewlett-Packard, Burroughs, and Prime Computer are rushing to buy applications software companies. Other

manufacturers, including Honeywell, Sperry, and Digital Equipment, are setting up joint ventures with software suppliers' (*ibid.*, p.55).

41. We have already seen how the US and Japanese governments have intervened and continue to intervene in supporting the development of their respective electronics industries. Similar efforts on a national basis are being implemented in various European and other countries of the world. For a description of government policies in different countries, see UNIDO RCSB (1985) and Hazewindus (1982).

42. In Scandinavia, Swedish and Finnish electronics companies have now merged to create a single and powerful Nordic Company (*Electronics,* 9 February 1984). Likewise, in Switzerland, Hasler AG and Autophon AG, the country's biggest communications houses, have agreed to merge their operations into a new company, ASCOM, which will produce about two-thirds of Swiss telecommunications equipment (*Electronics,* 8 January 1987). Another, intra-European merger involves the French electronics group Thomson and the Italian microelectronics concern SGS. The merger of their microchip activities will create Europe's second largest semiconductor company after Philips (*Financial Times*, 23 March 1987). In addition, the British General Electric Company and Philips have also entered into agreement merging their medical electronics businesses in a joint venture, and the Swedish company Ericsson has gained control of CGCT the French public exchange manufacturer (*Financial Times,* 9 July 1987).

43. This latest move has taken place mainly in the application-specific integrated circuits area where twenty-seven agreements were signed in 1984. As Iversen reports, 'chip vendors are teaming up to grab larger shares of the still fast-growing market for application-specific integrated circuits . . . The latest ASIC deal is a big one: Motorola Inc. and NCR Corp. have signed a pact that could make them a strong entry in gate arrays and standard cells . . . By trading computer-aided-design tools developed within each company and working together on design automation, the companies hope to make Motorola and NCR appear as "mirror images" to the customer' (Iversen, 1985, p.20). The most recent agreement in ASIC has brought together two of the largest US semiconductor manufacturers, TI and Intel, who are aiming to become leaders in this growing market (*Financial Times*, 8 July 1987).

44. In the sector of semiconductor production equipment, US companies are also entering into production agreements with Japanese companies as the focus of the Japanese market is shifting to local equipment makers. 'US suppliers are trying to cope with this shift without becoming dislodged, and they see the joint venture as the best means to accomplish this. Consequently, the number of joint ventures has increased at a rapid rate, with many of them formed only within the last 18 months' (Rothschild, 1983b, p.6).

45. Honeywell, one of America's biggest and oldest computer companies, is teaming with Japan's NEC in its attempt to hold its market share against IBM (*Business Week*, 7 November 1984).

46. TI has reached basic agreement with Canon Inc. (Japan) 'on joint development of various high technology products. Both companies hope to secure an advantageous position on the world's intensifying high technology market' (*Chemical Economy and Engineering Review,* March 1985, p.49). In addition, TI has technological agreements with the Dutch company Philips and the pan-European semiconductor group, European Silicon Structure (ESSI). These agreements will ensure the adoption of a common approach in their design methods of semi-custom chips, one of the fastest growing sectors of the semiconductor market (*Financial Times,* 15 December 1986).

Notes to Chapter 6

1. Various authors have written on the wide applicability and pervasive role of microtechnology throughout the technical realm of society. See, for instance, Evans (1979), Barron and Curnow (1979), Burkitt and Williams (1980), Forester (1980), Marsh (1981), Bessant *et al.* (1981), Bessant and Dickson (1982), Friedrichs and Schaff (1982). Molina (1987) provides an interpretation of the technical foundations for the pervasive nature of microtechnology, in historical perspective.

2. A number of studies have sought to show the existence of alternative technical options to the current development or implementation of microtechnology and the existence of very specific social options behind what is taking place. For instance, Noble (1979, 1984), Braverman (1974), Kaplinsky (1984) have dealt with the issue within the context of factory automation and, particularly, in relation to NC machine tools. Also, a number of CSE's (Conference of Socialist Economists) studies have touched upon such areas as robots, NC, the office, etc., [see CSE Microelectronics Group (1980), Duncan (1982)]. In the same vein, Walton (1982) has suggested the need to include human implications at the level of design, whereas E. Mumford *et al.* (1978), E. Mumford and Henshall (1979) and E. Mumford and Weir (1979) have proposed a participatory, sociotechnical approach to the design of work systems involving computers. Finally, Cooley (1980a, 1980b) and Rosenbrock (1977a, 1977b) have discussed the problem of social options in the development of technology in relation to the computerisation of design work. Among the studies reporting on experimental work aimed at establishing the viability of human-centred technical options, see Rosenbrock's and E. Mumford's references above, Dickson (1984), Howard (1985), Noble (1984), Athanasiou (1985). Finally, because of its importance as a social struggle aiming at redirecting modern technology into socially useful products and human-centred alternatives, the experience of the Lucas Combine — although not directly related to microtechnology — is highly relevant to this issue [see Wainwright and Elliot (1982), Cooley (1980a)].

3. Note that I am saying that this is the only *dominant* social constituent and not the only social constituent, for labour is also rooted in the microtechnology, insofar as it is itself a basic resource of

the technological process. The difference lies in the fact that labour is not part of the dominant social constituency of the technology. For a detailed treatment of the role of different social forces in technological processes, see Molina (1987).

4. In this context, science is used in its more common definition, namely, the pursuit of systematic knowledge of natural phenomena (Mulkay, 1977). According to Freeman (1974), 'the expression "*science-related*" technology is usually preferable to the expression "*science-based*" technology with its implication of an over-simplified one-way movement of ideas' (Freeman, p.29). On the other hand, the expression 'science-based' technology emphasises more clearly the fact that scientific knowledge is indeed fundamental to the existence of the technology.

5. Electricity and magnetism, thought of as separate phenomena, had already attracted the attention of the ancient Greeks. But according to Mason (1952), their study in modern times 'may be said to have begun with the researches of William Gilbert of Colchester during the sixteenth century' (Mason, p.474). On the other hand, Atherton (1984) relates the beginning of modern electrical science to the discovery of conduction by Stephen Gray in 1729. However, the science of electromagnetism as such may be said to have started only in 1820 with Hans Christian Oersted's announcement of the existence of the unified phenomenon of electromagnetism. Thereafter, during the nineteenth century, other great names contributed to establish the foundations of this science, among others André Marie Ampere with his mathematical work on electrodynamics (1820-2), George Ohm with his work on the relationship between current, voltage and resistance (1826-7), Michael Faraday with his work on electrolysis and electromagnetic induction (1831) and James C. Maxwell with his mathematical formulation of the theory of electromagnetism, first published between 1855 and 1864 [Dunsheath (1962), Atherton (1984)]. Later, in 1877, Heinrich Hertz proved the validity of Maxwell's theory while simultaneously building a rudimentary transmitter and detector which demonstrated the basis of communications across space. In the early-twentieth century Max Planck postulated the principles of quantum theory (1900), and Einstein in 1905 suggested the dual particle/wave nature of light and other electromagnetic radiation. In 1924, Louis de Broglie extended this concept and suggested that all matter has dual wave/particle properties. De Broglie's work was furthered by other scientists, notably, Werner Heisenberg, Erwin Schrodinger and Paul Dirac, and, in 1926, Schrodinger gave mathematical formulation to the wave behaviour of the electron [Mason (1962), Atherton (1984), Bernal (1969)]. In this way, scientists gradually developed the knowledge base leading to the understanding of matter and energy and hence, to the understanding of the semiconducting phenomenon at the base of microtechnology.

6. 'In that year Thomas Alva Edison, in his effort to increase the life of his early carbon-filament lamps, introduced a metal electrode into the vacuum envelope containing the glowing filament. He discovered that

when a positive voltage was applied to the electrode, a current flowed across the vacuum between it and the filament. This phenomenon — the great inventor's only fundamental scientific discovery — is the basis of all electron tubes and of all electronics up to the solid-state era' (*Electronics,* 1980, p.60).

7. See Gibbons and Johnson (1970), Braun and MacDonald (1977), MacDonald *et al.* (1981), Nelson (1962), Brooks (1976), Weiner (1979).

8. In 1839, Alexander Becquerel discovered that he could generate a voltage by illuminating the junction of an electrolyte. A century later, the latter was called a semiconductor. In 1873, Willoughby Smith discovered photoconductivity by observing the reduction in the resistance of illuminated selenium. Finally, in 1874, Ferdinand Braun discovered the rectifying property of contacts between metals and semiconductors [Gibbons and Johnson (1970), *Electronics* (1980)].

9. Wilson put forward a bond theory of conduction in semiconductors and explained the existence of both positive and negative highly mobile charge carriers (holes and electrons) as well as the effect of impurities on conductivity (Gibbons and Johnson, 1970).

10. For instance, with the rise of science-based technologies under capitalism, science has become a major factor of the sociotechnical and productive system governed by capital, so much so, that various scholars, particularly within the Marxist school of thought, have postulated that science has become a direct productive force [Svorykin (1963), ISA (1977), Hales (1982)] and has been pressed into the service of capital interests [Braverman (1974), Noble (1977), Rose and Rose (1976a), Albury and Schwartz (1982)]. Most authors see this process as being a twentieth-century development although it would have started by the late nineteenth century with the emergence of the science-based chemical and electrical industries. Marx himself, however, wrote in both *Capital* and *Grundrisse* about the pressing of science into the service of capital with the development of modern industry. For Braverman (1974), however, in Marx's own days 'this was . . . more an anticipatory and prophetic insight than a description of reality' (Braverman, p.155). On the other hand, Cooper (1971) argues that in referring to science, Marx's 'terminology is designed to cover a new kind of economically oriented activity involving the search for new inventions, and the rational scientific examination of production processes made possible by the factory system' (Cooper, pp.178-9). See also Rosenberg (1974) for Marx's concept of science.

11. For Marx (1867), the essence of the process of capital accumulation is the self-expansion of value through the generation of surplus value, i.e., the increment of excess over the original value advanced by the capitalist. 'Capital is, by definition, any value that is increased by surplus value' (Mandel, 1977, p.81). In Marx's own words, 'The value originally advanced . . . not only remains intact while in circulation, but adds to itself a surplus-value or expands itself. It is this movement that converts it into capital' (Marx, p.149).

12. 'Capital arises from surplus-value. Employing surplus-value as capital, reconverting it into capital, is called accumulation of capital' (Marx, (1867), p.543).

13. 'This is the specific difference between centralization and concentration, the latter being only another name for reproduction on an extended scale. Centralization may result from a mere change in the distribution of capitals already existing, from a simple alteration in the quantitative grouping of the component parts of social capital. Here capital can grow into powerful masses in a single hand because then it has been withdrawn from many individual hands . . . Centralization completes the work of accumulation by enabling industrial capitalists to extend the scale of their operations' (Marx, (1867) pp.587-8).

14. 'The further machine production advances, the higher becomes the organic composition of capital needed for entrepreneurs to secure the average profit. The average capital needed in order to start a new enterprise capable of bringing in this average profit increases in the same proportion. It follows that the average size of enterprises likewise increases in every branch of industry. Those enterprises will be the most likely to succeed in competition which have an organic composition of capital which is above the average, which possess the largest reserves and funds for most rapidly advancing along the road of technical progress' (Mandel, 1977, p.163).

15. This development has been a conspicuous aspect of the electronics industry insofar as the technology itself has created the conditions for convergence and integration. Thus, big industries are tending to penetrate new fields continuously as well as integrate backwards and forward from raw materials to market outlets.

16. This process which Rada (1982) has called the increasing dependence of production on capital is in fact a major underlying feature of the drive towards automation commonly associated with the development of microtechnology. In addition, technical change in the case of microelectronics is leading to capital-saving techniques too.

17. In this connection, the conglomerate corporation is the ultimate expression of centralisation. According to Dos Santos (1972), these conglomerates 'operate in an enormous number of economic sectors having no technological link among themselves' (Dos Santos, p.19). On the other hand, in relation to the science-based industries concerning microtechnology, we can see from the Appendix that, from their early roots in the communications industry, a strong oligopolistic tendency has been at work.

18. It is commonplace to identify the Manhattan Project and its development of the first atomic bomb as the critical event which best epitomises the nature of the social complex of interests and power formed in World War II. In Hales' words, in this project 'large numbers of workers of high intellectual standing were employed, together with massive resources of industrial production like those needed to refine uranium ores. Scientists, directors and managers of industrial corporations, and

military and civilian state officials were all forced to work closely by the urgency and secrecy of the project but also by the intrinsic complexity of the apparatus which spanned fields of work that were normally separate' (Hales, 1982, p.109).

Note to Chapter 7

1. 'US agricultural employment collapsed from 20 percent of the labour force in 1929 to 3 percent 50 years later' (*Financial Times*, 2 May 1987, p.16).

Notes to Appendix

1. As Recabarren (1980) put it, 'The emergence of monopolies was a natural result of the dynamics of concentration and centralization intrinsic to the process of capital accumulation', but, at the same time, in a dialectical fashion, 'with the development of monopolies an acceleration of the very same process that was its cause, that is to say, capital accumulation, had to take place' (Recabarren, p.19).

2. 'Imperialism is capitalism in that stage of development in which the dominance of monopolies and finance capital has established itself; in which the export of capital has acquired pronounced importance; in which the division of all territories of the globe among the great capitalist powers has been completed' (Lenin, 1944, p.77). In its most succinct definition, for Lenin, 'imperialism is the monopoly stage of capitalism' (*ibid.*).

3. The corporation, under the name of joint-stock company, had begun its development in Marx's time. Among its consequences, Marx saw in the corporation an 'enormous expansion of the scale of production and of enterprises, that was impossible for individual capitals', and also a 'transformation of the actually functioning capitalist into a mere manager, administrator of other people's capital, and of the owner of capital into a mere owner, a mere money capitalist' (Marx, 1894, p.436).

4. 'The big corporation came into its own in the second half of the nineteenth century, first in the fields of finance and railways spreading to industry around the turn of the century, and later invading most other branches of the national economy' (Baran and Sweezy, 1975, p.40). See also Braverman (1974).

5. For instance, 'Between 1896 and 1905 the size of the hundred largest American companies quadrupled and by 1905 they controlled 40 percent of America's industrial capital' (Friedman, 1977, p.38).

6. In his explanation of the transformation of industry by the late nineteenth century, Hobsbawm (1978) argues that among the most important changes, the 'first and in the long run most profound change was in the role of science in technology . . . The major technical advances of the second half of the nineteenth century were . . . essentially scientific . . . Two major growth industries of the new phase of industrialism, the electrical and the chemical, were entirely based on scientific knowledge

. . . The last major change was the increase in the *scale* of economic enterprise, the concentration of production and ownership, the rise of an economy comprised of a handful of great lumps of rock-trusts, monopolies, oligopolies — rather than a large number of pebbles' (Hobsbawm, pp.172 and 177).

7. The first specialised technical schools emerged in France and were organised by the absolutist state 'strictly to meet the needs of transportation, mining, the military and the Navy' (Bohme *et al.*, 1978, p.227). In this sense, as Weingart suggests, they 'seem to have been the result of an early utilization of science by the state' (Weingart, 1978, p.270). Among some of the early schools in France were, the *École des Ponts et Chaussées* (1750), the *École du Corps des Ingénieurs des Mines* (1778), the *École Royale Militaire* (1753), the *École du Corps Royal du Génie* (1765). In England, similar institutions were the Mechanics Institutes, and in Germany the Mining Academies in Berlin (1775), in Freiberg (1765), and the Schools of Agronomy (Bohme *et al.*, 1978). In France, the movement of technical schools led in 1794 to the formation of the *École Polytechnique* which, according to Drucker (1961), marked the establishment of the profession of engineer. The main focus of these schools was technical training, but they incorporated the fundamentals of mathematics and natural science in their curriculum. Hence, Weingart's statement that 'they were not limited to a purely technical training function but also fostered natural sciences' (Weingart, 1975, p.270).

8. According to Hales (1982), 'The systematic application of research effort to problems relevant to commercial practice took off most notably in Germany, where the chemistry teaching laboratory of Justus von Liebig (founded at Giessen 1824) became a model of future development' (Hales, p.89).

9. Braverman (1974) points out that the leadership in chemistry and its industrial applications first belonged to France and was forged during the Napoleonic Wars as a result of the cutting of supplies of soda, sugar and other products. In Bernal's view, the birthplace of chemical research was eighteenth-century Britain. But starting with Lavoisier, France gained and kept a seventy-year supremacy in chemical research. Later, the leadership slipped to Germany (Bernal, 1969).

10. As Bernal (1944) has stated, 'in a few years the chemistry of dye-stuffs and explosives, for which the foundation had been laid largely in France and Britain, had been captured as part of a new German industry which held the virtual monopoly of the world market' (Bernal, p.29).

11. 'The six largest German firms for coal-tar products took out 948 patents between 1886 and 1990, as compared with 86 by the corresponding English firms' (Landes, 1969, pp.352-3). Also, 'the six largest German chemical works employed more than 650 chemists and engineers, while the entire British coal-tar industry had no more than thirty or forty' (Braverman, 1974, p.162). See also Barber (1970).

12. 'Whatever the situation in the early nineteenth century, formal links between science and industry were increasingly forged in the form of a distinct organisation within the firm — the research and development laboratory. Moreover, the specialisation of inventive and innovative activity was not restricted to modern, science-related industries; it was also found in established industries like iron and steel . . . it became increasingly dangerous for firms to be left behind by technical change and in turn more profitable for them to invest in the exploration of new products and processes' (Pavitt and Worboys, 1977, p.17). In the same vein, Noble defines modern science-based industry as 'industrial enterprise in which ongoing scientific investigation and the systematic application of scientific knowledge to the process of commodity production have become routine parts of the operation' (Noble, 1977, p.5).

13. Although not engaged in fundamental research, Edison's laboratory made full use of the fundamental laws of electromagnetism. Thus, 'Having . . . defined the purpose, Edison achieved it through the systematic application of the scientific discoveries which had been made by Ohm, Oersted, Laplace, Joule, Faraday and others' (Sabato, 1975, p.39). In his *Networks of Power*, Hughes has dealt extensively with the characteristics of Edison's laboratory. According to him, 'The Edison laboratory at Menlo Park was probably one of the best electrical laboratories in the world. Moreover, Edison also equipped it, at great expense, as a chemical research laboratory . . . Edison assembled a community of craftsmen and appliers of science and the tools and scientific instruments they needed in order to work on problems of a systematic nature' (Hughes, 1983, pp.24-5).

14. For a brief historical review of early laboratories in different industrial fields, see Barlett (1941) and Fleming (1917).

15. 'A few people feared that technological improvement might be hampered by a lack of scientific information. Some leaders at General Electric at the end of the century feared that the industry was rapidly "using up" its scientific capital. The organisation of the General Electric Research Laboratory was one step in an attempt to rectify that situation' (Birr, 1979, p.197).

16. For instance, in reference to GE's decision to set up the Schenectady laboratory in 1900, Lindsay says 'It was soon recognized by the directors of this new company that the amount of technological development which could be drawn out of the scientific knowledge already accumulated during the nineteenth century, though large, was finite and that there would be a greater chance of ingenious developments if there were more science to work with' (Lindsay, 1973, p.216).

17. In one scholar's description, 'the industrial research laboratory . . . grew up almost imperceptibly from the workshop or private testing place of the inventor turned businessman, such as Siemens or Edison' (Bernal, 1969, p.569). The names of Alexander Bell and Guglielmo Marconi are of particular relevance for the present study of electronics as they were

the inventor-entrepreneurs who did most in relation to the telephone and radio respectively.

18. 'Nearly all of the basic research done by industry, as well as the bulk of applied research, was restricted to large firms with ample financial resources, since they alone were able to provide researchers with a relatively stable working situation and adequate facilities' (Noble, 1977, p.111). Also, 'only corporations of great wealth . . . can afford large research organizations. Good industrial research of any kind, and specially relatively fundamental research, is expensive for a number of reasons. In the first place, good research talent and good research facilities are not cheap . . . In the second place, research is expensive because there may be a period of anywhere from five to ten years between the original scientific conception or "hunch" and its application in an actual industrial process or product . . . And, last of all, the development work which lies between pure research and industrial application is also very costly, not only in equipment but in engineering talent' (Barber, 1970, pp.220-1).

19. In the telephone industry, for instance, the 'Bell System's permanent commitment to research came in 1907 with the consolidation of its research activities in the Western Electric Company and AT&T, and was established institutionally in 1911 with the creation of its first research branch' (Hoddeson, 1981, p.515). This was not a sudden development, however, since from the establishment of the Engineering Department in 1881, the Bell company had been organising departments to deal with the technical problems imposed by the growing telephone industry. Thus, what took place in 1907 was in fact a major reorganisation of the various existing facilities, bringing them together under a greater commitment to research [Fagen (1975), Coon (1939), Hoddeson (1981)].

20. 'On the whole those industries born in the laboratory or directly dependent upon new knowledge for their growth organized research activities earlier and more rapidly than the industries which had long been established. In fact in 1920 approximately two-thirds of all the research workers who were recorded in the first survey of the National Research Council were employed in the electrical, chemical, and rubber industries' (Barlett, 1944, p.34).

21. Both the American Bell Telephone Co. and Marconi in radio fought fiercely to control their respective markets. This situation was reflected in the technological and R&D policies of both companies. At Bell's, for instance, the establishment of the Engineering Department in 1881 was particularly the result of a strategy to control the market through an extension of the company's patent production. The Department was 'to conduct research and experimentation and to evaluate outside inventions for relevance to the telephone' (Brocks, 1981, p.103). For this reason Coon (1939) suggested that 'The Bell System is built on patents. Its objective is the perpetuation of its monopoly' (Coon, p.7). Another example within Bell was the company's decision to develop long-distance lines which was in practice an important aspect of the strategy 'to develop industrial

control' (Brocks, 1981, p.104). On the other hand, in the field of radio, as McClaurin (1949) described, 'Marconi's plans for marine wireless were large and ambitious. He hoped to control the basic patents in the art, and to equip ships of all nations with wireless apparatus. He hoped also to erect shore stations at key points around the world, through which all ships' messages would be sent. In the pursuit of these objectives, Marconi was determined to obtain a monopolistic position' (McClaurin, p.37).

22. A particular instance described by Hoddeson (1981) relates to the influence on R&D of the Bell System's dream of a universal system by constructing intercontinental lines. In her words, 'Out of Vail's [president of the Bell System at the time, 1908-9] largely nonscientific goal to create a universal telephone system grew his decision to build a transcontinental line, from which came the technological problem of developing a non-mechanical repeater, and this problem in turn contributed crucially to the start of Bell's formal commitment to in-house basic research. "Basic" industrial research was now recognized as intrinsically dual in nature, being fundamental from the point of view of the researchers while at the same time supported by the company for its possible applications' (Hoddeson, p.534-5).

23. According to Sherwood (1967) a patent proposal may have been advanced by Hippodamus already in Ancient Greece, but 'the earliest patent law on record was enacted in 1474 by the Republic of Venice' (Sherwood, p.488). In England, the patent system dates from the sixteenth century and a statute was passed by Parliament in 1623. It was geared towards the promotion of new industries by granting monopolies to importers of inventions as well as to inventors themselves. In the US, the first Patent Law was enacted in 1790 and empowered Congress 'to promote the Progress of Science and useful Arts, by securing for limited times for Authors and Inventors the exclusive right to their respective Writings and Discoveries' (quoted by Sherwood, 1967, pp.488-9). Since then patents have been considered to protect and reward the inventor, i.e., 'to be either a reward to the inventor or a result of a bargain between him and society' (Bernal, 1944, p.144). By 1860, this may have indeed been the case when Abraham Lincoln praised the virtues of the patent system for adding 'the fuel of interest to the fire of genius' (quoted by Noble, 1977, p.84). But with the rise of big corporations and the industrial laboratory seeking to control the process of technological change, the situation has changed entirely. In Noble's words, 'Within a half-century later after Abraham Lincoln offered his glowing evaluation of it, the American patent system has undergone a dramatic change; rather than promoting invention through protection of the inventor, the patent system had come to protect and reward the monopolizer of inventors, the science-based industrial corporations' (*ibid.*, p.85).

24. In the 1930s, for instance, a report of the National Resource Committee claimed that the Bell Telephone System had suppressed 3,400 unused patents in order to forestall competition (Sherwood, 1967).

25. An illuminating example of this is what happened at the end of the Bell System's patent monopoly of the telephone in 1894. Prior to this for seventeen years Bell had enjoyed a total monopoly of the telephone market and the annual rate of return on investment had been approximately forty-six percent (Brock, 1981). After 1894, without the patent monopoly, the situation changed markedly. 'By 1900 telephone competition was widespread. The independents controlled 38 percent of the phones installed in the United States . . .[and] . . . Return on investment for the system as a whole declined . . . to 8 percent during the years 1900-1906' (*ibid.*, p.117).

26. Consider, for instance, how the invention of the vacuum tube by Lee de Forest and the development of radio became major technical challenges to the Bell System's standing in the telephone industry [Reich (1977), Coon (1939), Brock (1981)]. Commercial and legal moves 'protected AT&T from existing competitors using the existing technology, it did not protect the company against major technological change. Technological change could threaten either by allowing another company to enter existing types of telephone service or by creating a new product that would take business away from the telephone. Radio and the vacuum tube threatened both kinds of changes. Because the vacuum tube was crucial to a telephone amplifier for long-distance service, a company that controlled the vacuum tube could provide severe long-distance competition. Because radio could communicate without wires, it could lead to a new product to displace wire-based telephone systems' (Brock, 1981, p.175). A policy of systematic R&D helped AT&T to contain both challenges. First, it enabled the company swiftly to spot the strategic importance of the vacuum tube and eventually to acquire its patent from de Forest. Second, it enabled AT&T to forestall any invasion of the telephone field by getting involved in the radio field first and thus negotiating from a position of strength the spheres of influence to be left to competing companies. As Reich (1977) put it, 'the use of patent rights could take both an offensive and a defensive character, and the defense could be indirect' (Reich, p.209-10).

27. As Purcell (1966) has commented, 'there were certain functions which were either clearly unprofitable or impossible for industry to provide for itself' (Purcell, p.232).

28. The influence of industry on university science was also noted by Bernal (1969), 'university laboratories also grew, from the very fact that the new uses of science meant new jobs and attracted more and more students. Thus, despite all the protestations and disinterestedness, the academic science of the period was ultimately dependent on the success of science in industry' (Bernal, p.569).

29. For instance, see Barber (1970), Barlett (1941), Noble (1977), Brand (1941), Miller (1966), Coben (1979).

30. The First World War has been characterised as the 'Chemists' War' [Lasby (1966), Rose and Rose (1977)].

31. 'From the standpoint of industrial research . . . the obvious connection between Germany's scientific establishment and its military prowess, combined with great shortages of commodities such as dyes and pharmaceuticals following the outbreak of the war in 1914, went far to convince many influential Americans of the necessity to give organised science strong support' (Lewis, 1967, p.628).

32. In this connection, see Pursell (1966), Lindsay (1973), Lasby (1966), Penick *et al.* (1972), Mark (1982), Pavitt and Worboys (1977).

33. Another important effect of the First World War upon science was a severe blow on scientists' traditional values as they embarked in the cause of war. 'The tradition of scientists' neutrality and internationalism was rudely shaken, if not totally destroyed by the outbreak of the First World War. Scientists and technologists in all the combatant countries responded to nationalistic propaganda and their own patriotic promptings by enlisting their talents in the service of the nation-state' (Lakoff, 1977, p.358).

34. The impact of this situation for the case of AT&T's research, for instance, has been described as follows: 'From 1917 to 1918 long-distance telephony research was discontinued. Efforts in the war years centered on the development of two-way radio telephone sets for dispatch purposes on subchasers and airplanes. In 1919 AT&T resumed its former program' (McClaurin, 1949, p.93).

35. This is not to suggest that antagonism between the diverse social interests actually developed. After all, all of them shared a common interest in the expansion of the R&D system and, as we know, the main overriding concern of capital is not necessarily to produce for the civilian market but to ensure the reproduction of the profit-driven process of capital accumulation. In this respect, the military market may well offer important advantages in terms of profits and monopolistic protection, thus enticing industrial interests into heavy military involvement. On the other hand, government may well have strong interests in military power as part of its national and international politics, while the overriding concern of scientific interests seems to lie much more with the reproduction and expansion of research and research facilities than with any supposedly anti-militaristic ethical principle. Finally, it is also true that scientific and technological progress for military purposes does not exclude attaining civilian technological benefits, since many advances may be of generic nature with application in both military and civilian fields.

36. 'Some useful work continued, especially within the Naval Research Laboratory, the Ordnance and Signal Corps, and the National Advisory Committee for Aeronautics, but the results were not such as to elicit enthusiasm' (Lasby, 1966, p.263). See also Swain (1967).

37. During the war, the US government confiscated all 4,500 German chemical patents and used them as a base for creating a strong US chemical industry. The rationale of such a policy was not simply that it would provide an immediate solution to the wartime shortage of chemicals but that it 'would also serve to protect the new industry against German

competition after the war' (Barlett, 1941, p.36). Although not in such a dramatic fashion, the creation of a strong US radio industry was achieved by similar government intervention in the face of major competition by the British and German radio industry [McClaurin (1949), Reich, (1977), Freeman (1974)]. In the case of radio, spurred by commercial and strategic considerations, the US government and the Navy promoted the creation of a unified American company which would give the country a powerful and autonomous standing in the field of radio. To this end, GE was persuaded to buy a controlling interest in American Marconi — the subsidiary of the British company controlling the world market in the aftermath of the war — and in 1919, the Radio Corporation of America (RCA) was established. In addition, in 1920 a pooling of the radio patents in possession of all the companies with big interests in radio (i.e., GE, AT&T, and Westinghouse) was arranged. We may say, therefore, that in the cases of the chemical and the radio industries, a convergence of government, military and industrial interests gave these industries a powerful social constituency which profoundly shaped their development in the national and international context.

38. A similar realisation of the importance of R&D for industrial competitiveness also influenced the British, as is shown by a report of the Department of Scientific and Industrial Research published in 1932. There it was argued that as a result of the war, 'there was a general awakening to the fact that for success in times of peace as well as of war it was desirable that the resources of science should be utilized to the full. The perils of war furnished the precepts for peace, and it was realised that on the conclusion of the conflict a situation would arise in the world of industry which would call for increased effort if British industrial supremacy was to be maintained, and if the manufactured products of the nation were to continue to hold their own in the world's markets' (Bernal, 1944, pp.30-1).

39. Robert Millikan and George Hale had led the efforts to revitalise the US National Academy of Science (NAS) taking advantage of the opportunity offered by the First World War. When the National Research Council (NRC) was formed in 1916 as a subsidiary of the NAS, to centralise the scientific war effort, Millikan became chief executive officer and 'the focal point of a large scientific enterprise' (Swain, 1967, p.538). The NRC, with its numerous committees, subcommittees and specialised panels, 'became the vehicle through which hundreds of individual scientists could participate in the wartime science program. The National Research Council solicited the co-operation of both university scientists and industrial specialists to carry out its mission. Moreover, it won the co-operation of the great foundations who showered their largesse on the NRC when federal funds proved inadequate' (*ibid.*).

40. The post-war decline in the role of the government social constituent was greatly deepened by the impact of the Great Depression of the 1930s. 'The Great Depression of the 1930s had a disastrous

impact on government-supported science activity. The downward spiral of appropriations for research forced virtually every scientific agency of the federal government to retrench' (Swain, 1967, p.539).

41. For Dupree (1957) the impact of the war had such a decisive effect that, in his view, 'industrial research as a branch of the country's scientific establishment dates its rise to eminence almost entirely from the war period' (Dupree, p.323).

42. It is interesting to note that of the total research personnel reported in 1940, slightly more than half (52.2%) were professionally trained, mainly as chemists and engineers. The remaining part was about equally divided between technical (23.4%) and non-technical (e.g. administrative, maintenance) workers (24.4%) (Cooper, 1941).

SELECT BIBLIOGRAPHY

The following are some of the best analyses exemplifying the general tenets of the social-shaping-of-technology approach.

Noble, D. *America by Design: Science and Technology and the Rise of Corporate Capitalism,* A. Knopf, New York, 1977.

Hughes, T. *Networks of Power: Electrification in Western Society 1880-1930,* Johns Hopkins University Press, Baltimore, 1983.

MacKenzie, D. and Wajcman, J. (eds.), *The Social Shaping of Technology,* Open University Press, Milton Keynes, England, 1985.

Winner, L. *The Whale and the Reactor: A Search for Limits in an Age of High Technology,* University of Chicago Press, Chicago, 1986.

For the particular case of the social shaping of microtechnology:

CSE Microelectronics Group. *Capitalist Technology and the Working Class,* CSE Books, London, 1980.

Noble, D. *Forces of Production,* A. Knopf, New York, 1984.

Kaplinsky, R. *Automation. The Technology and Society*, Longmans, London, 1984.

Blackburn, P., Coombs, R. and Green, K. *Technology, Economic Growth and the Labour Process,* MacMillan, London, 1985.

The role of the military in the development of microtechnology is dealt with in the following collections of essays.

Tirman, J. *The Militarization of High Technology,* Ballinger, Camb., Mass., 1984.

Smith, M.R. (ed.), *Military Enterprise and Technological Change: Perspectives on the American Experience,* MIT Press, Camb., Mass., 1985.

An overview of the great variety of issues related to the microelectronics revolution is found in the following collections of essays.

Forester, T. *The Microelectronics Revolution,* Basil Blackwell, Oxford, 1980.

Forester, T. *The Information Technology Revolution,* Basil Blackwell, Oxford, 1985.

A few books have dealt with issues, primarily economic issues, of microelectronics and the Third World.

Rada, J. *The Impact of Micro-Electronics,* International Labour Organization, Geneva, 1980.

Ernst, D. *Global Race in Microelectronics: Innovation and Corporate Strategy in a Period of Crisis,* Campus, Frankfurt, 1983.

Jacobsson, S. and Sigurdson, J. (eds.), *Technological Trends and Challenges in Electronics: Dominance of the Industrialized World and Responses in the Third World.* Research Policy Institute, Univ. of Lund, Sweden, 1983.

GENERAL BIBLIOGRAPHY

Abegglen, J. and Etori, E. 'Japanese Technology Today' (Special Report), *Scientific American*, Vol. 249, No. 4, (1983), pp.J-1 to J-24.

Abernathy, W. 'The Competitive Decline in US Innovation: The Management Factor' in H. Fusfeld and R. Langlois (eds.), *Understanding R&D Productivity*, Pergamon Press, New York, 1982, pp.55-72.

Adam, J. and Fischetti, M. 'SDI: The Grand Experiment', *IEEE Spectrum*, September 1985, pp.34-5.

Adam, J. and Horgan, J. 'Debating the Issues', *IEEE Spectrum*, September 1985, pp.55-64.

Adam, J. and Wallich, P. 'Mind-Boggling Complexity', *IEEE Spectrum*, September 1985, pp.36-46.

Adams, G. *The Politics of Defense Contracting: The Iron Triangle*. Transaction Books, London, 1982.

Adams, G. and Gold, D. 'Recasting the Military Spending Debate'. *Bulletin of the Atomic Scientists*, October 1986, pp.26-32.

Aftergood, S. 'Nuclear Space Mishaps and Star Wars', *Bulletin of the Atomic Scientists*, October 1986, pp.40-3.

Albury, D. and Schwartz, J. *Partial Progress: The Politics of Science and Technology*, Pluto Press, London, 1982.

Allison, D. *The Origins of Radar at the Naval Research Laboratory: A Case Study of Mission-Oriented Research and Development*, Ph.D. Thesis, Princeton University, 1980.

Anderson, H. 'IBM Versus Bell in Telecommunications', *Datamation*, May 1977, pp.91-5.

Athanasiou, T. 'High-Tech Alternativism: The Case of the Community Memory Project', *Radical Science*, No. 16, 1985, pp.37-51.

Atherton, W. *From Compass to Computer*, San Francisco Press, San Francisco, 1984.

Aviation Week & Space Technology. 'Technical Survey: Very High Speed Integrated Circuits', 16 February 1981, pp.48-85.

Baran, P. and Sweezy, P. *Monopoly Capital*, Penguin Books, Harmondsworth, Middlesex, England, 1985.

Barber, B. *Science and the Social Order*, Collier Books, New York, 1970.

Barlett, H. 'The Development of Industrial Research in America' in National Resource Planning Board (ed.) (1941), pp.19-77.

Barnes, B. and Edge, D. 'The Interaction of Science and Technology' in B. Barnes and D. Edge (eds.). *Science in Context*, Open University Press, Milton Keynes, England, 1982, pp.147-54.

Barron, I. and Curnow, R. *The Future with Microelectronics: Forecasting the Effects of Information Technology*, Frances Pinter, London, 1979.

Battelle-Colombus Division, *Probable Levels of R& D Expenditures in 1985: Forecast and Analysis,* Battelle, Ohio, 1984.

Bell, T. 'The Decision to Divest: Incredible or Inevitable?', *IEEE Spectrum*, November 1985, pp.46-55.

Beresford, R. 'VHSIC: Redefining the Mission', *VLSI Design,* November 1983, pp.16-21.

Berger, M. 'Japanese Firms Boost Spending for Short- and Long-Term Projects', *Electronics Week,* 24 September 1984, pp.32-6.

Berger, M. 'Why US Wants Military Technology from Japan', *Electronics,* 29 July 1985, pp.34-5.

Bernal, J. *The Social Function of Science,* George Routledge, London, 1944.

Bernal, J. *Science in History: The Natural Science in Our Time (Vol 3),* Penguin Books, Harmondsworth, Middlesex, England, 1969.

Bessant, J. *Technology and Market Trends in the Production and Application of Information Technology: A Review of Developments During the Years 1982-1983.* UNIDO/IS.438, Vienna, 1984.

Bessant, J. and Dickson, K. *Issues in the Adoption of Microelectronics,* Frances Pinter, London, 1982.

Bessant, J., Bowen, J., Dickson, K. and Marsh, J. *The Impact of Microelectronics: A Review of the Literature,* Frances Pinter, London, 1981.

Birr, K. 'Science in American Industry' in D. van Tassel and M. Hall (eds.) 1966, pp.35-80.

Birr, K. 'Industrial Research Laboratories', in N. Reingold (ed.) (1979), pp.193-207.

Blackburn, P., Coombs, R. and Green, K. *Technology, Economic Growth and the Labour Process,* MacMillan, London, 1985.

Bohme, G., van Den Daele, W. and Krohn, W. 'The "Scientification" of Technology' in W. Krohn *et al.*, (eds.) (1978), pp.219-50.

Borrus, M., Millstein, J. and Zysman, J. *International Competition in Advanced Industrial Sectors: Trade and Development in the Semiconductor Industry,* US Government Printing Office, Wash., DC, 1982.

Botkin, J. and Dimancescu, D. 'The DARPA Exception' in J. Tirman (ed.) (1984), pp.222-5.

Botkin, J., Dimancescu, D. and Stata, R. *Global Stakes: The Future of High Technology in America,* Ballinger Pub. Co., Camb., Mass., 1982.

Brand, C. 'Technical Research by Trade Associations' in National Resources Planning Board (ed.) (1941), pp.88-97.

Braun, E. 'From Transistors to Microprocessors' in T. Forester (ed.) (1980), pp.72-102.

Braun, E. and MacDonald, S. *Revolution in Miniature: The History and Impact of Semiconductor Electronics,* Cambridge University Press, Cambridge, 1978.

Braverman, H. *Labor and Monopoly Capital: The Degradation of Work in the Twentieth Century,* Monthly Review, New York, 1974.

Brock, G. *The US Computer Industry: A Study of Market Power,* Ballinger Pub. Co., Camb., Mass., 1975.

Brock, G. *The Telecommunications Industry: The Dynamics of Market Structure,* Harvard University Press, Camb., Mass., 1981.

Brooks, J. *Telephone: The First Hundred Years,* Harper and Row, New York, 1976.

Brown, H. 'Star Wars Once Funny, Now Frightening', *Bulletin of the Atomic Scientists,* Vol. 41, No. 5, 1985, p3.

Bueche, A. 'The Economy' in H. Fusfeld and C. Haklisch (eds.) (1979), pp. 138-53.

Burke, J. and Eakin, M. (eds.), *Technology and Change,* Boyd and Fraser, San Francisco, 1979.

Burkitt, A. and Williams, E. *The Silicon Civilization,* W. Allen, London, 1980.

Bylinsky, G. 'The Holes in AT&T's Computer Strategy', *Fortune,* 13 September 1984, pp.50-62.

Cain, S. and Adam G. 'Reagan's 1988 Military Budget', *Bulletin of the Atomic Scientists,* March 1987, pp.50-52.

Chomsky, N. 'The Cornerstone of the American System' in Y. Fitt, A. Faire and J-P Vigier (1980), pp.1-12 (foreword).

Clery, D. 'SDI Technology — too far, too fast?', *Electronics and Power,* July 1987, pp. 437-42.

Coben, S. 'American Foundations As Patrons of Science: The Commitment to Individual Research' in N. Reingold (ed.) (1979), pp.229-47.

Connolly, R. 'Pentagon to Fund Major IC Program', *Electronics,* 14 September 1978, pp.81-2.

Cooley, M. *Architect or Bee: The Human/Technology Relationship,* Langley Technical Series, Slough, England, 1980a.

Cooley, M. 'The Designer in the 1980s — The Deskiller Deskilled', *Design Studies,* Vol 1, No. 4, 1980b, pp.197-201.

Coon, H. *American Tel & Tel: The Story of a Great Monopoly,* Longmans, New York. Reprinted in 1981, University Microfilms International, London.

Cooper, C. 'Science, Technology and Development', *Economic and Social Review* (Dublin), Vol. 2, No. 2, 1971, pp.165-89.

Cooper, F. 'Location and Extent of the Industrial Research Activity in the United States' in National Resources Planning Board (1941) pp.173-87.

Cortes-Comeres, N. 'A Venerable Giant Sharpens Its Claws', *IEEE Spectrum,* February 1986, pp.54-65.

CSE Microelectronics Group, *Capitalist Technology and the Working Class,* CSE Books, London, 1980.

David, E. 'Science Futures: The Industrial Connection' *Science,* Vol. 203, No. 4383, 1974, pp.837-40.

De Grasse, R. 'The Military and Semiconductors' in J. Tirman (ed.) (1974) pp.77-104.

de Sola Pool, I. (ed.) *The Social Impact of the Telephone,* MIT Press, Camb., Mass., 1977.

de Sola Pool, I. *Technologies of Freedom,* Harvard University Press, Camb., Mass., 1983.

Dickson, D. *The New Politics of Science,* Pantheon Books, New York, 1984.

Diebold, J. *Automation: The Advent of the Automatic Factory,* D. van Nostrand Co., New York, 1952.

Diebold, J. *Beyond Automation: Managerial Problems of an Exploding Technology,* McGraw-Hill Book Co., New York, 1964.

Dinneen, G. and Frick, F. 'Electronics and National Defense', *Science,* Vol. 195, No. 4283, 1987, pp.1151-5.

Dos Santos, T. 'Las Contradicciones del Imperialismo Contemporaneo', *Sociedad y Desarrollo,* Enero/Marzo (1972), pp.9-34.

Drucker, P. 'The Technological Revolution: Notes on the Relationship of Technology, Science and Culture', *Technology and Culture,* Vol. 2, No. 4, 1961, pp.342-51.

Dumas, L. 'University Research, Industrial Innovation, and the Pentagon' in J. Tirman (ed.) (1984), pp.123-51.

Dumas, L. 'The Military Burden on the Economy', *Bulletin of the Atomic Scientists,* October 1986, pp.22-5.

Duncan, M. 'The Information Technology Industry in 1981', *Capital and Class,* No. 17, 1981, pp.78-113.

Dunsheath, P. *A History of Electrical Engineering,* Faber and Faber, London, 1962.

Dupree, A. *Science in the Federal Government,* Harvard University Press, Camb., Mass., 1957.

Electronics. Special Commemorative Issue on the History of Electronics, April 1980.

Ellul, J. 'The Biology of Technique', *The Nation,* 24 May 1965, p.568.

Ellul, J. *The Technological Society,* Alfred Knopf, New York, 1967.

Ellul, J. *Hope in Time of Abandonment,* The Seabury Press, New York, 1972.

Emme, E. 'The Contemporary Spectrum of War' in M. Kranzberg and C. Pursell (eds.) (1967), pp.578-90.

Enfield, R. 'The Limits of Software Reliability', *Technology Review*, Vol. 90, No. 3, 1987, pp.36-43.

Ernst, D. *Restructuring World Industry in a Period of Crisis: The Role of Innovation.* UNIDO/IS.285, Vienna, 1981. Later published in 1983 as *The Global Race in Microelectronics. Innovation and Corporate Strategy in a Period of Crisis,* Campus, Frankfurt.

Evans, C. *The Mighty Micro: The Impact of the Computer Revolution,* Victor Gollancz, London, 1979.

Fagen, M. (ed.) *A History of Engineering and Science in the Bell System: The Early Years (1875-1925),* Bell Laboratories, USA, 1975.

Farber, D. and Baran, P. 'The Convergence of Computing and Telecommunications Systems', *Science,* Vol. 195, No. 4283, 1977, pp.1166-70.

Federal Register, 'Initial Military Critical Technologies List', Vol. 45, No. 192, 1980, pp.65014-9.

Feigenbaum, E. and McCorduck, P. *The Fifth Generation: Artificial Intelligence and Japan's Computer Challenge to the World,* Pan Books, London, 1983.

Fels, R. *The Second Crisis of Economic Theory,* General Learning Press, New Jersey, 1972.

Ferguson, C. 'The Microelectronics Industry in Distress', *Technology Review,* Vol. 86, No. 6, 1983, pp.24-37.

Finan, W. 'The Semiconductor Industry's Record on Productivity' in SIA, *American Prosperity and Productivity: Three Essays on the Semiconductor Industry,* SIA, Cupertino, Calif., 1981, pp.19-28.

Finn, B. 'Electronics Communications' in M. Kranzberg and C. Pursell (eds.) (1967), pp.293-309.

Fischetti, M. 'Exotic Weaponry', *IEEE Spectrum,* September 1985, pp.47-54.

Fischetti, M. 'A Review of Progress at MCC', *IEEE Spectrum,* March 1986, pp.76-82.

Fisher, F., McKie, J. and Mancke, R. *IBM and the US Data Processing Industry: An Economic History,* Praeger, New York, 1983.

Fishman, K. *The Computer Establishment,* McGraw-Hill Book Co., New York, 1981.

Fitt, Y. 'Two Tools for Domination: Industry and Agriculture' in Y. Fitt, A. Faire and J-P Vigier (1980), pp.13-67.

Fitt, Y., Faire, A. and Vigier, J-P. *The World Economic Crisis: US Imperialism at Bay,* Zed Press, London, 1980.

Fleck, G. *A Computer Perspective,* Harvard University Press, Camb., Mass., 1973.

Fleming, A. *Industrial Research in the United States of America,* HMSO, England, 1917.

Foremski, T. 'AT&T: Can It Do Battle With IBM', *Computing,* 12 January 1984, p.16.

Forester, T. *The Microelectronics Revolution,* Basil Blackwell, Oxford, 1980.

Forester, T. *The Information Technology Revolution,* Basil Blackwell, Oxford, 1985.

Freeman, C. *The Economics of Industrial Innovation,* Penguin, Harmondsworth, Middlesex, England, 1974.

Freeman, C. 'Economics of Research and Development' in I. Spiegel-Rosing and D. de Solla Price (eds.) (1977), pp.223-75.

Freeman, C. 'The Economic Implications of Microelectronics' in C. Cohen (ed.) (1982), *Agenda for Britain 1: Micro Policy, Choices for the 1980s*, Phillip Allan, Oxford, 1982, pp.53-88.

Freeman, C. 'The Economics of Innovation', *IEE Proceedings,* Vol. 132, Pt. A, No. 4, July 1985, pp.213-21.

Freeman, C., Clark, J. and Soete, L. *Unemployment and Technical Innovation: A Study of Long Waves and Economic Development,* Frances Pinter, London, 1982.

Friedrichs, C. and Schaff, A. (eds.) *Microelectronics and Society: For Better or For Worse,* Pergamon Press, Oxford, 1982.

Fusfeld, H. and Haklisch, C. *Science and Technology Policy: Perspectives for the 1980s*, The New York Academy of Sciences, New York, 1979.

Galbraith, J. *The New Industrial State,* Penguin Books, Harmondsworth, Middlesex, England, 1978.

Galbraith, J. *Economics and the Public Purpose,* Andre Deutsch, London, 1974.

Gansler, J. *The Defense Industry,* MIT Press, Camb., Mass., 1980.

Gibbons, M. and Johnson, C. 'Relationship Between Science and Technology', *Nature,* Vol. 227, 11 July 1970, pp.125-7.

Gibbons, M. and Wittrock, B. (eds.), *Science as a Commodity: Threats to the Open Community of Scholars,* Longman, England, 1985.

Golding, A. *The Semiconductor Industry in Britain and the United States: A Case-Study in Innovation, Growth and Diffusion of Technology,* D. Phil. Dissertation, University of Sussex, 1971.

Goldman, M. 'The Shifting Balance of World Power', *Technology Review*, Vol. 90, No. 3, 1987, pp.20-21.

Goldstine, H. *The Computer From Pascal to Von Neumann,* Princeton University Press, New Jersey, 1972.

Gregory, G. 'Japan Challenges the Computer Giant', *New Scientist,* 6 January 1983, pp.28-30.

Hales, M. *Science or Society? The Politics of the Work of Scientists,* Pan Books, London, 1982.

Hazewindus, N. *The US Microelectronics Industry: Technical Change, Industry Growth and Social Impact,* Pergamon Press, New York, 1982.

Hobsbawm, E. *Industry and Empire*, Penguin Books, Harmondsworth, Middlesex, England, 1978.

Hoddeson, L. 'The Emergence of Basic Research in the Bell Telephone System, 1875-1915', *Technology and Culture,* Vol. 22, No. 3, 1981, pp.512-44.

Hogan, L. 'Reflections on the Past and Thoughts About the Future of Semiconductor Technology', *Interface Age,* Vol. 2, No. 4, 1977, pp.24-36.

Howard, R. 'UTOPIA: Where Workers Craft New Technology', *Technology Review,* Vol. 88, No. 3, 1985, pp.43-9.

Hudson, C. 'Computers in Manufacturing', *Science,* Vol. 215, No. 4534, 1982, pp.818-25.

Hughes, T. *Networks of Power: Electrification in Western Society 1880-1930,* John Hopkins University Press, Baltimore, 1983.

Inaba, M. 'Japan Mfrs. Seek Tech. Feedback via US Sales', *Electronics News,* 7 March 1983 (Supplement), pp.10 and 22-3.

Irwin, M. and Johnson, S. 'The Information Economy and Public Policy', *Science,* Vol. 195, No. 4283, 1977, pp.1170-74.

ISA (International Sociological Association) collection of papers. *Scientific-Technological Revolution: Social Aspects,* Sage Pub., London, 1977.

Iversen, W. 'VHSIC-Insertion Program Begins to Pay Dividends', *Electronics Week,* 19 July 1984, pp.57-66.

Iversen, W. 'Motorola and NCR Team Up in Semicustoms Chips', *Electronics,* 29 July 1985, pp.20-21.

Jacobsson J. *Problems and Issues Concerning the Transfer, Application and Development of Technology in the Capital Goods and Industrial Machinery Sector,* UNCTAD, TD/B/C.6/AC713, 1982.

Johnson, J. 'Strategies in the Services Realm', *Datamation*, July 1982, pp.24-30.

Kaldor, M. 'The Armament Process' in D. MacKenzie and J. Wajcman (eds.) (1985), pp.263-9.

Kaplinsky, R. *Computer-Aided Design: Electronics, Comparative Advantage and Development,* Frances Pinter, London, 1982.

Kaplinsky, R. *Automation: The Technology and Society,* Longman, London, 1984.

Katz, B. and Phillips, A. 'The Computer Industry' in R. Nelson (ed.) (1982), pp.162-232.

Kikuchi, M. *Japanese Electronics,* Simul International, Tokyo, 1983.

Kilby, J. 'Invention of the Integrated Circuit', *IEEE Transactions on Electron Devices,* Vol. ED-23, No. 7, 1976, pp.648-54.

Kirkland, R. 'Ma Blue: IBM's Move Into Communications', *Fortune,* 15 October 1984, pp.36-40.

Kozma, R. 'AT&T's Global Plan to Score in Computer Market', *Electronics,* 5 August 1985, pp.26-9.

Kranzberg, M. and Pursell, C. (eds.), *Technology in Western Civilization,* Vol. 2, Oxford University Press, London, 1967.

Krohn, W., Layton, E. and Weingart, P. (eds.), *The Dynamics of Science and Technology,* D. Reidel Pub. Co., Dordrecht, Holland, 1978.

Kulish, V. 'Science and Warfare', *Impact of Science on Society*, Vol. 26, No. 1/2, 1976, pp.53-61.

Kurth, J. 'The Political Economy of Weapons Procurement: The Follow-On Imperative' in R. Fels, (ed.) (1972), pp.304-11.

Lakoff, S. 'Scientists, Technologists and Political Power' in I. Spiegel-Rosing and D. de Solla Price (eds.) (1977), pp.355-91.

Lamb, J. 'Star Wars' Software Will Not Work', *New Scientist,* 5 September 1985, p.19.

Lamborghini, B. 'The Impact on the Enterprise' in G. Friedrichs and A. Schaff (eds.) (1982), pp. 119-56.

Landes, D. *The Unbound Prometheous,* Cambridge University Press, Cambridge, 1969.

Lasby, C. 'Science and the Military' in D. van Tassel and M. Hall (eds.) (1966), pp.251-82.

Layton, E. 'Conditions of Technological Development' in I. Spiegel-Rosing and D. de Solla Price (eds.) (1977), pp.197-222.

Lenin, V. *Imperialism, the Highest Stage of Capitalism,* Lawrence and Wishart, London, 1944.

Levin, R. 'The Semiconductor Industry' in R. Nelson (ed.) (1982), pp.9-100.

Levin, R. 'R&D Productivity in the Semiconductor Industry: Is a Slowdown Imminent?' in H. Fusfeld and R. Langlois (eds.). *Understanding R&D Productivity,* Pergamon Press, New York, 1982, pp.37-54.

Lewis, W. 'Industrial Research and Development' in M. Kranzberg and C. Pursell (eds.) (1967), pp.615-34.

Lin, H. 'The Software for Star Wars: An Achilles Heel?', *Technology Review,* Vol. 88, No. 5, 1985, pp. 16-18.

Lindsay, R. *The Role of Science in Civilization,* Greenwood Press, Westport, Connecticut, 1973.

Lineback, R. 'Ford Sees Silicon Through Colorado Facility', *Electronics,* 26 January 1984, p.106.

Linvill, J. and Hogan, L. 'Intellectual and Economic Fuel for the Electronics Revolution', *Science,* Vol. 195, No. 4283, 1977, pp.1107-13.

Louis, A. 'The Great Electronic Mail Shootout', *Fortune,* 20 August 1984, pp.149-52.

MacDonald, S., Collingridge, D. and Braun E. 'From Science to Technology: The Case of Semiconductors', *Bull. Sci. Tech. Soc.,* Vol. 1, No. 1-2, 1981, pp.173-201 and No. 3 pp.289-320.

MacKenzie, D. *Behind the News: The Military, the Computer and Artificial Intelligence,* Sociology Department, University of Edinburgh, April 1985.

MacKenzie, D. and Wajcman, J. (eds.), *The Social Shaping of Technology,* Open University Press, Milton Keynes, England, 1985.

Mandel, E. *Marxist Economic Theory,* Merlin Press, London, 1977.

Mansfield, E. *The Economics of Technological Change,* Longmans, London, 1969.

Mansfield, E. *Research and Innovation: The Modern Corporation,* MacMillan, London, 1972.

Marbarch, W. *et al.,* 'The Race to Build a Supercomputer' in T. Forester (ed.) (1985), pp.60-70.

Mark, H. 'Technology and the Strategic Balance', *Technology in Society,* Vol. 4, No. 1, 1982, pp.15-32.

Marsh, P. *The Silicon Chip Book,* Abacus, London, 1981.

Marsh, P. 'The Heroic Obstacles Still to Be Overcome', *Financial Times,* 25 June 1985, p.16.

Marshall, E. 'William Perry and the Weapons Gamble', *Science,* Vol. 211, No. 4483, 1981, pp.681-3.

Marx, C. *Early Writings,* Penguin Books, Harmondsworth, Middlesex, England, 1975.

Marx, C. *Capital,* Vol. 1, Lawrence and Wishart, London, 1977a. (First published in German in 1867).

Marx, C. *Capital,* Vol. III, Lawrence and Wishart, London, 1977b. (First published in German in 1894).

Marx, C. *Grundrisse,* Penguin Books, Harmondsworth, Middlesex, England, 1977c. (First published in German in 1939).

Mason, S. *A History of the Sciences,*Collier Books, New York, 1962.

Mason, J. 'VLSI Goes to School', *IEEE Spectrum,* November 1980, pp.48-52.

McCartney, L. 'Exxon: Another Computer Giant?', *Datamation,* July 1978, pp.169-73.

McClaurin, R. *Invention and Innovation in the Radio Industry,* The MacMillan Co., New York, 1949.

Melman, S. 'Swords Into Plowshares: Converting from Military to

Civilian Production', *Technology Review*, Vol. 89, No. 1, 1986, pp.62-71.

Merrifield, D. 'Forces of Change Affecting High Technology Industries', *National Journal,* 29 January 1983, pp.253-6.

Miller, H. 'Science and Private Agencies' in D. van Tassel and M. Hall (eds.) (1966), pp.191-221.

Misa, T. 'Military Needs, Commercial Realities, and the Development of the Transistor, 1948-1958' in M. R. Smith (ed.). *Military Enterprise and Technological Change: Perspectives on the American Experience,* MIT Press, Camb., Mass., 1985.

Molina, A. H. 'The US Revalues its Electronics Patents', *New Scientist,* 1 May 1986, pp.40-43.

Molina, A. H. *The Sociotechnical Basis of the Microelectronics Revolution: A Global Perspective,* Ph.D. Thesis, 2 Vols, University of Edinburgh, 1987.

Molina, A. H. 'Managing Economic and Technological Competitiveness in the US Semiconductor Industry: Short- and Long-Term Strategies', *International Journal of Technology Management,* forthcoming.

Mowery, D. 'Innovation, Market Structure, and Government Policy in the American Semiconductor Electronics Industry: A Survey', *Research Policy*, Vol. 12, No. 4, 1983, pp.183-97.

Mulkay, M. 'Sociology of the Scientific Research Community' in I. Spiegel-Rosing and D. de Solla Price (eds.) (1977), pp.93-148.

Mumford, E. and Henshall, D. *A Participative Approach to Computer Systems Design: A Case Study of the Introduction of a New Computer System,* Associated Business Press, London, 1979.

Mumford, E., Land, F. and Hawgood, J. 'A Participative Approach to the Design of Computer Systems', *Impact of Science on Society,* Vol. 28, No. 3, 1978, pp.235-53.

Mumford, E. and Weir, M. *Computer Systems in Work Design — the ETHICS Method: Effective Technical and Human Implementation of Computer Systems,* Associated Business Press, London, 1979.

Mumford, L. *Technics and Civilization,* Routledge and Kegan Paul, London, 1967.

Mumford, L. *The Myth of the Machine: The Pentagon of Power,* Harcourt Brace Jovanovich, New York, 1970.

Naegele, T. 'GE-RCA — A New Powerhouse or a Stodgy Behemoth', *Electronics,* 6 January, 1986, pp.73-5.

NAE/NRC (National Academy of Engineering/National Research Council). *The Competitive Status of the US Machine Tool Industry,* National Academy Press, Wash., DC, 1983.

NAE/NRC. *The Competitive Status of the US Electronics Industry,* National Academy Press, Wash., DC, 1984.

National Resource Planning Board. *Research — A National Resource,* Vol. II, (1941), NRPB, Wash., DC.

National Science Foundation. *Research and Development in Industry, 1970,* NSF, Wash., DC, 1971.

National Science Foundation (NSF). *Research and Development in Industry, 1982,* NSF, Wash., DC, 1984.

Nelkin, D. and Nelson, R. 'University-Industry Alliances', *Science, Technology and Human Values,* Vol. 12, No. 1, 1987, pp.65-74.

Nelson, R. 'The Link Between Science and Invention: The Case of the Transistor' in National Bureau of Economic Research. *The Rate and Direction of Inventive Activity,* Princeton University Press, Princeton, 1962, pp.549-83.

Nelson, R. 'Technical Advance and Economic Growth: Present Problems and Policy Issues' in H. Fusfeld and C. Haklisch (eds.) (1979), pp.47-57.

Nelson, R. (ed.). *Government and Technical Progress: A Cross-Industry Analysis,* Pergamon Press, New York, 1982.

Niosi, J. and Faucher, P. 'The Decline of North American Industry', *IDS Bulletin,* Vol. 16, No. 2, 1985, pp.12-19.

Noble, D. *America by Design: Science, Technology, and the Rise of Corporate Capitalism,* Alfred Knopf, New York, 1977.

Noble, D. 'Social Choice in Machine Design: The Case of Automatically Controlled Machine Tools' in A. Zimbalist (ed.). *Case Studies on the Labour Process.* Monthly Review, New York, 1979, pp.18-50.

Noble, D. 'The Selling of the University', *The Nation,* 6 February 1982, pp.129 and 143-8.

Noble, D. *Forces of Production,* Alfred Knopf, New York, 1984.

Noble, D. 'Present Tense Technology' in Collective Design Projects (eds.). *Very Nice Work If You Can Get It: The Socially Useful Production Debate,* Spokesman, Nottingham, England, 1985. First published during 1983 in *Democracy: A Journal of Political Renewal and Radical Change,* New York.

Norman, C. *Knowledge and Power: The Global Research and Development Budget,* Worldwatch Paper 31, Wash., DC, 1979.

Norman, C. *The God That Limps: Science and Technology in the Eighties,* W. Norton, New York, 1981.

Norman, C. 'Electronics Firms Plug Into the Universities', *Science,* Vol. 217, No. 4559, 1982, pp.511-14.

OECD. *Reviews of National Science Policy: United States,* OECD, Paris, 1968.

OECD. *Gaps in Technology: Electronics Components,* OECD, Paris, 1968b.

OECD. *Science, Growth and Society,* OECD, Paris, 1971.

OECD. *The Measurement of Scientific and Technical Activities: 'Frascati Manual',* OECD, Paris, 1976.

Ornstein, S., Smith, B. and Suchman, L. 'Strategic Computing', *Bulletin of the Atomic Scientists,* Vol. 40, No. 10, 1984, pp.11-15.

Panofsky, W. 'The Strategic Defense Initiative: Perception vs Reality', *Physics Today,* June 1985, pp.34-35.

Pappademos, J. 'Militarized Science and the Crisis of Capitalism Today', *Science and Nature,* No. 6, 1983, pp.6-20.

Patel, K. and Bloembergen, N. 'Strategic Defense and Directed-Energy Weapons', *Scientific American,* Vol. 257, No. 3, 1987, pp.31-37.

Pavitt, K. and Worboys, M. *Science, Technology and the Modern Industrial State,* Butterworth, London, 1977.

Peck, M. 'Joint R&D: The Case of Microelectronics and Computer Technology Corporation', *Research Policy*, Vol. 15, 1986, pp.219-31.

Penick, J., Pursell, C., Sherwood, M. and Swain, D. *The Politics of American Science: 1939 to the Present,* MIT Press, Camb., Mass., 1972.

Perez, C. 'Microelectronics, Long Waves and World Structural Change: New Perspectives for Developing Countries', *World Development,* Vol. 13, No. 3, 1985, pp.441-63.

Perry, W. and Roberts, G. 'Winning Through Sophistication: How to Meet the Soviet Military Challenge', *Technology Review*, July 1982, pp.27-35.

Platzek, R. and Kilby, J. 'Minuteman Integrated Circuits: A Study of Combined Operations', *IEEE Proceedings,* Vol. 52, 1964, pp.1669-78.

Ploch, M. 'Industry Invests in Research Centers', *High Technology,* May 1983, pp.15-18.

Prager, D. and Ommen, G. 'Research, Innovation, and University-Industry Linkages', *Science,* Vol. 207, No. 4429, 1980, pp.379-84.

Price, D. *The Scientific Estate,* Oxford University Press, Oxford, 1965.

Pugh, E. *Memories That Shaped an Industry: Decisions Leading to IBM System/360,* MIT Press, Camb., Mass., 1984.

Pursell, C. 'Science and Government Agencies' in D. van Tassel and M. Hall (eds.) (1966), pp.223-49.

Rada, J. 'Microelectronics, Information Technology and Its Effects on Developing Countries' in J. Berting, S. Mills and H. Wintersberger (eds.). *The Socioeconomic Impact of Microelectronics,* Pergamon Press, Oxford, 1980.

Rada, J. 'Technology and the North-South Division of Labour,' *IDS Bulletin,* Vol. 13, No. 2, 1982, pp.5-13.

Rae, J. *Climb to Greatness: The American Aircraft Industry 1920-1960,* MIT Press, Camb., Mass., 1968.

Recabarren, J. *Latin America: Dependence and Trade Unionism,* M. Phil Dissertation, University of Bradford, 1980.

Reich, M. 'Does the US Economy Require Military Spending' in R. Fels (ed.) (1972), pp.296-303.

Reich, L. 'Research, Patents, and the Struggle to Control Radio: A Study of Big Business and the Use of Industrial Research', *Business History Review,* No. 51, 1977, pp.208-35.

Reingold, N. (ed.). *The Sciences in the American Context: New Perspectives,* Smithsonian Institution Press, Wash., DC, 1979.

Reppy, J. 'The United States' in N. Ball and M. Leitenberg (eds.). *The Structure of the Defense Industry,* Croom Helm, London 1983, pp.21-49.

Robinson, A. 'Giant Corporations From Tiny Chips Grow', *Science,* Vol. 208, No. 4443, 1980a, pp.480-84.

Robinson, A. 'Perilous Times for US Microcircuit Makers', *Science,* Vol. 208, No. 4444, 1980b, pp.582-6.

Rose, H. and Rose, S. *The Political Economy of Science,* MacMillan, London, 1976.

Rose, H. and Rose, S. 'The Production of Science in Advanced Capitalist Society' in H. Rose and S. Rose (eds.) (1976), pp.14-31.

Rose, H. and Rose, S. *Science and Society,* Penguin Books, Harmondsworth, Middlesex, England, 1969.

Rosenberg, N. 'Karl Marx on the Economic Role of Science', *Journal of Political Economy,* Vol. 82, No. 4, 1974, pp.56-77.

Rosenberg, N. 'The Economic Implications of the VLSI Revolution', *Futures,* October 1980, pp.358-69.

Rosenberg, N. *Inside the Black Box: Technology and Economics,* Cambridge University Press, Cambridge, 1983.

Rosenbrock, H. 'Interactive Computing: A New Opportunity', *Control Systems Center Report No. 388,* University of Manchester (UMIST), September 1977a.

Rosenbrock, H. 'The Future of Control', *Automatica,* Vol. 13, 1977b, pp.389-92.

Rothschild, K. 'US, Japan Equipment Mfrs. Vie for R&D Edge', *Electronics News*, 7 March 1983a (Supplement), pp.4-5.

Rothschild, K. 'US Suppliers Enter Production Deals in Japan', *Electronics News,* 7 March 1983b (Supplement), pp.6, 18-19, 26.

Sabato, J. 'Using Science to "Manufacture" Technology', *Impact of Science on Society*, Vol. 25, No. 1, 1975, pp.37-44.

Salomon, J-J. 'Science Policy Studies and the Development of Science Policy' in I. Spiegel-Rosing and D. de Solla Price (eds.) (1977), pp.443-71.

Salomon, J-J. 'Science as A Commodity — Policy Changes, Issues and Threats' in M. Gibbons and B. Wittrock (eds.) (1985), pp.78-98.

Sapolsky, H. 'Science, Technology and Military Policy' in I. Spiegel-Rosing and D. de Solla Price (eds.) (1977), pp.443-71.

Schnee, J. 'Government Programs and the Growth of High-Technology Industries', *Research Policy,* No. 7, 1978, pp.2-24.

Schlesinger, T. 'Labour, Automation and Regional Development' in J. Tirman (ed.) (1984a), pp.181-213.

Schmitt, R. 'National R&D Policy: An Industrial Perspective', *Science,* Vol. 224, No. 4654, 1984, pp. 1206-9.

Schroeder-Gudehus, B. 'Science, Technology and Foreign Policy' in I. Spiegel-Rosing and D. de Solla Price (eds.) (1977), pp.473-506.

Schroeer, D. *Physics and Its Fifth Dimension: Society*, Addison-Wesley Pub. Co., London, 1972.

Sciberras, E. 'The UK Semiconductor Industry' in K. Pavitt (ed.). *Technical Innovation and British Economic Performance,* MacMillan, London, 1980, pp.282-96.

Sciberras, E. and Payne, B. *Technical Change and International Competitiveness: A Study of the Machine Tool Industry,* The Technical Change Centre, London, 1984.

Science 'Special Issue on Japanese Technology and Industry', Vol. 231, No. 4741, 1986.

Science for People. 'Star Wars Special Issue', No. 61, Summer 1986.

Servan-Schreiber, J-J. *The World Challenge.* Simon and Schuster, New York, 1980.

Shapley, D. 'Electronics Industry Takes to "Potting" Its Products for Market', *Science,* Vol. 202, No. 4370, 1978, pp.848-9.

Sharpe, W.*The Economics of Computers,* Columbia University Press, New York, 1969.

Sherwood, M. 'Technology and Public Policy' In M. Kranzberg and C. Pursell (eds.) (1967), pp.487-98.

Shulman, S. 'Stopping Star Wars', *Science for People,* Vol. 18, No. 1, 1986, pp.10-15.

SIA (Semiconductor Industry Association). *The International Microelectronic Challenge: The American Response by the Industry, the Universities and the Government*, SIA, Cupertino, California, 1981.

Sigurdson, J. *Japan's High Technology Race: The Information Technologies,* RPI, Lund, Sweden, 1983.

Silk, L. *The Research Revolution,* McGraw-Hill, New York, 1960.

SIPRI, 'Gasto y Producción Militares en el Mundo', *Comercio Exterior,* Vol. 35, No. 3, 1985, pp.268-81.

Smith, J. 'Soviet Lag in Key Weapons Technology', *Science,* Vol. 219, No. 4590, 1983, pp.1300-1.

Smith, M. (ed.). *Military Enterprise and Technological Change: Perspectives on the American Experience,* MIT Press, Camb., Mass., 1985.

Soma, J. *The Computer Industry: An Economic-Legal Analysis of Its Technology and Growth*, Lexington Books, Mass., 1978.

Soukup, M. 'The Scientific-Technical Revolution and a Comprehensive Disarmament Programme', *Impact of Science on Society*, Vol. 26, No. 1-2, 1976, pp.91-9.

Spiegel-Rosing, I. and de Solla Price, D. (eds.). *Science, Technology and Society: A Cross-Disciplinary Perspective,* Sage Pub., London, 1977.

Staat, E. 'The Economy: Introductory Remarks' in H. Fusfeld and C. Haklisch (eds.) (1979), pp.135-7.

Stanfield, R. 'Campuses and Corporations: Industry Offers Money, But Not Without Strings', *National Journal*, 29 November 1980, pp.2021-4.

Sturmey, S. *The Economic Development of Radio,* Gerald Duckworth, London, 1958.

Sumney, L. 'VLSI with a Vengeance', *IEEE Spectrum*, April 1980, pp.24-7.

Sun, M. 'The Pentagon's Ambitious Computer Plan', *Science,* Vol. 222, No. 4629, 1983, pp.1213-15.

Sundaram, G. 'Is Military R&D a Necessary Evil?', *Impact of Science on Society,* Vol. 31, No. 1, 1981, pp.5-13.

Svorykin, A. 'Science as A Direct Productive Force', *Impact of Science on Society,* Vol. 13, No. 1, 1963, pp.49-60.

Swain, D. 'Organization of Military Research' in M. Kranzberg and C. Pursell (eds.) (1967), pp.535-48.

Sweezy, P. *The Theory of Capitalist Development,* Monthly Review, London, 1970.

Sweezy, P. *Modern Capitalism and Other Essays,* Monthly Review, New York, 1972.

Teal, G. 'Single Crystals of Germanium and Silicon —Basic to the Transistor and the Integrated Circuit', *IEEE Transaction on Electron Devices,* Vol. ED-23, No. 7, 1976, pp.621-9.

Thee, M. 'Significancy of Military R&D: The Impact of the Arms Race on Society', *Impact of Science on Society*, Vol. 31, No. 1-2, 1981, pp.49-59.

Thurow, L. 'A World-Class Economy: Getting Back Into the Ring', *Technology Review,* Vol. 88, No. 6, 1985, pp.27-37.

Tilton, J. *International Diffusion of Technology: The Case of Semiconductors,* The Brookings Institution, Wash., DC, 1971.

Tirman, J. 'The Defense-Economy Debate' in J. Tirman (ed.) (1984), pp.1-32.

Tirman, J. *The Militarization of High Technology,* Ballinger, Camb., Mass., 1984.

Tucker, J. 'The Strategic Computer Initiative', *Science for the People,* Vol. 17, No. 1-2, 1985, pp.21-5.

Tugendhat, C. *The Multinationals,* Penguin Books, Harmondsworth, Middlesex, England, 1981.

UNIDO RCSB (Regional and Country Studies Branch). *Survey of Government Policies in Informatics,* UNIDO/IS.526, Vienna, 1985.

US Department of Commerce. Bureau of Census. *Historical Statistics of the United States: Colonial Times to 1970,* Part 2, Wash., DC, 1975.

US Department of Commerce. Industry and Trade Administration. *A Report on the US Semiconductor Industry*, September 1979, Wash., DC.

Uttal, B. 'Japan's Latest Assault on Chipmaking', *Fortune,* 3 September 1984, pp.58-65.

Utterback, J. and Murray, A. *Influence of Defense Procurement and Sponsorship of Research and Development on the Development of the Civilian Electronics Industry,* Center for Policy Alternatives, MIT Press, Camb., Mass., June 1977.

Vandercook, W. 'SDI Show Hits the Road', *Bulletin of the Atomic Scientists,* Vol. 42, No. 8, 1986, pp.16-18.

van Tassel, D. and Hall, M. (eds.). *Science and Society in the United States,* The Dorsey Press, Illinois, 1966.

Vigier, J-P. 'The Crisis and the Third World War' in Y. Fitt, A. Faire and J-P. Vigier (1980), pp.111-77.

von Hippel, F. 'Attack on Star Wars Critic: A Diversion', *Bulletin of the Atomic Scientists,* Vol. 41, No. 4, 1985, pp.8-10.

Wainwright, H. and Elliot, D. *The Lucas Plan: A New Trade Unionism in the Making*, Allison and Busby, London, 1982.

Wallich, P. 'New Directions for a National Resource', *IEEE Spectrum*, November 1985, pp.90-96.

Wallich, P. 'US Semiconductor Industry: Getting It Together', *IEEE Spectrum*, April 1986, pp.75-8.

Walsh, J. 'International Trade in Electronics: US-Japan Competition', *Science,* Vol. 195, No. 4283, 1977, pp.1175-9.

Walsh, J. 'DoD Funds More Research in Universities', *Science*, Vol. 212, No. 4498, 1981, pp.1003-4.

Walsh, J. 'Japan-US Competition: Semiconductors Are the Key', *Science,* Vol. 215, No. 4534, 1982, pp.825-9.

Walton, R. 'Social Choice in the Development of Advanced Information Technology', *Technology in Society,* Vol. 4, No. 1, 1982, pp.41-49.

Webster, F. and Robin, K. 'Mass Communications and "Information Technology"', *Socialist Register,* 1979, pp.285-316.

Webster, F. and Robin, K. 'Technological Determination or Demystification?' *New Universities Quarterly*, Vol. 35, No. 3, 1981, pp.315-22.

Wedlake, F. *SOS: The Story of Radio Communications,* David and Charles, Newton Abbot, England, 1973.

Weinberger, C. *Annual Report to the Congress, Fiscal Year 1986,* Wash., DC, 1985.

Weiner, C. 'How the Transistor Emerged' in J. Burke and M. Eakin (eds.) 1979, pp.251-61.

Weingart, P. 'The Relation Between Science and Technology — A Sociological Explanation' in W. Krohn *et al.* (eds.) 1978, pp.251-86.

Weisberg, L. 'DoD Directions in Electronic Device R & D', *Proceedings of IEEE Conference on US Technological Policy,* 1978, pp.24-5.

Weinzenbaum, J. 'Computers in Uniform: A Good Fit?' *Science for the People,* Vol. 17, No. 1-2, 1985, pp.26-9.

White, L. 'Clearing the Legal Path to Cooperative Research', *Technology Review,* Vol. 38, No. 5, 1985, pp.38-44.

Whitington, F. 'Sizing Each Other Up', *Datamation*, July 1982, pp.8-16.

Wilson, R., Ashton, P. and Egan, T. *Innovation, Competition, and Government Policy in the Semiconductor Industry,* Lexington Books, Lexington, Mass., 1980.

Winner, L. *Autonomous Technology. Technics-Out-of-Control as a Theme in Political Thought,* MIT Press, Camb., Mass., 1977.

Winner, L. *The Whale and the Reactor. A Search for Limits in an Age of High Technology.* University of Chicago Press, Chicago, 1984.

Wohl, A. 'Office Automation: The Next Battlefield', *Datamation*, July 1982, pp.34-9.

Wolff, M. 'The Genesis of the Integrated Circuit', *IEEE Spectrum*, August 1976, pp.45-53.

Yonas, G. 'Strategic Defense Initiative: The Politics and Science of Weapons in Space', *Physics Today,* June 1985, pp.24-32.

Zinberg, D. 'The Legacy of Success: Changing Relationships in University-Based Scientific Research in the United States' in M. Gibbons and B. Wittrock (eds.) 1985, pp.107-27

Zorpette, G. 'The Telecommunications Bazaar', *IEEE Spectrum*, November 1985, pp.56-63.

INDEX